Electric Cars

by Brian Culp

Electric Cars For Dummies®

Published by: **John Wiley & Sons, Inc.,** 111 River Street, Hoboken, NJ 07030-5774, www.wiley.com

For general information on our other products and services, please contact our Customer Care Department within the U.S. at 877-762-2974, outside the U.S. at 317-572-3993, or fax 317-572-4002. For technical support, please visit https://hub.wiley.com/community/support/dummies.

Wiley publishes in a variety of print and electronic formats and by print-on-demand. Some material included with standard print versions of this book may not be included in e-books or in print-on-demand. If this book refers to media such as a CD or DVD that is not included in the version you purchased, you may download this material at http://booksupport.wiley.com. For more information about Wiley products, visit www.wiley.com.

Library of Congress Control Number: 2022942340

ISBN: 978-1-119-88735-5 (pbk); 978-1-119-88618-1 (ebk); 978-1-119-88619-8 (ebk)

SKY10035532_072822

Contents at a Glance

Table of Contents

Introduction

So you're thinking about buying an electric car?

Congrats!

You've come to the right place. Also, you've come to the right person, or at least I think you'll agree that's the case by the time you're done reading this intro. (Oh, and note to self: That first intro sentence might make a great chapter title.)

I won't hold it against you if you skip ahead right now to start getting answers to questions about vehicle range, charging, and cost of operation, especially when compared to internal combustion vehicles.

Don't worry. By the end of this book, you'll have explored almost every topic that someone new to electric cars might have on their mind. You'll get up close and personal with some fascinating engineering and future-looking topics, like why electric cars go so damn fast, why they tend to be much safer, what to expect in terms of battery longevity, and how and why the car you buy today is likely to improve over time.

Along with the topics covering the purchase and ownership experience — information especially relevant if you're new to the idea of electric cars — you can find a *wealth* of ancillary material about everything from autonomous driving, regen braking, electric trucks and bikes, all the way to suggestions about YouTubers that would be worth a follow as you continue down the electrification rabbit hole.

I'll even work in an anecdote about the Norwegian pop band A-ha and how their lead singer fits into humanity's movement toward electric transportation. Trust me: It'll make sense once we get there.

Most of all, here's what I want you to know about the person writing about the current landscape of electric vehicles — a confession that should provide some insight into your tour guide's brain.

I'm not really a car guy.

Instead, I'm more of an easier-way-to-get-from-point-A-to-B guy. I'm also an amazed-at-what-human-ingenuity-has-been-able-to-pull-off-and-we're-just-getting-started guy.

Here's what I mean: We humans are amazing creatures. Of course, plenty of said humans vex and frustrate on the daily, but to get a sense of what I'm talking about, simply look up the next time you're out for a drive.

When you pause to glance toward the horizon, you're likely to see majestic buildings of glass and rock and steel. You might see a metal tube hovering in midair, carrying people from one cluster of buildings to another cluster half a world away. You might see bridges spanning otherwise uncrossable bodies of water. You might notice towers carrying currents of electricity. These currents are capable of making toast, or lighting up the dark, or powering massive cars and trains. You might notice other towers that can send more information than a single human can read in a lifetime into the palm of your hand — all in a matter of seconds.

Amazing. More amazing still? We've made all these things by *digging them out of the mud*.

I'm a technical writer at Lucid Motors, a relatively new company that manufactures electric-only vehicles. My wife is a software engineer at Tesla Motors, another relatively new company that only makes electric vehicles (EVs, for short). We also drive these very same vehicles, having begun our journey of EV ownership starting in 2018. What I've realized over the past couple of years, however, is that the most important part of all the astounding things humans make — all the buildings and bridges and phones and planes — isn't the glass, steel, rubber, or silicon. Not by a long shot.

So, what is the most important part?

Information.

At its core, an electric vehicle — as it is with a laptop, suspension bridge, or life-saving medical implant — is an *idea*. All these inventions, along with everything else that humans have plucked out of the mud, have started with, and are perpetuated by, someone putting an idea on a page.

What makes my job so fulfilling is that I get to be part of a team that captures Lucid's idea for electric transportation. This idea is bigger and more enduring than any one man or woman. In fact, it's bigger than any one company. Lucid and Tesla and everyone else builds its electric vehicles by leveraging ideas that have been around for over a century.

The idea of an electric vehicle can therefore be revised, enhanced, or added to. Complementary technologies can converge for a multiplier effect. A more efficient motor can make use of a denser battery pack, which can increase car range and reduce costs, which can result in more EV purchases, which can incentivize further electric motor improvement — a true virtuous cycle of innovation.

Over time, the superfluous or useless can be cast aside while the better parts endure. Because the most important part of a good idea is that it *survives*, that the idea is passed from group to group, from generation to generation, and that those who succeed us will make the idea their own.

That's my mindset as we begin: I get to capture and present the *ideas* surrounding electric transportation as I think most directly relates to you, and I couldn't be more excited about this opportunity. I'll do the best I can to present the same kinds of information I've been exposed to over the past several years that have left me, a non-car guy, so enthusiastic about the future of electric mobility, and in turn cautiously optimistic about the future of humankind.

So buckle up, adjust the seats and mirrors, and prepare for an exhilarating ride as we take the idea of electric vehicles out for a spin.

About This Book

Books in the *For Dummies* brand are organized in a modular, easy-to-access format that lets you use the book more or less the way you would an owner's manual for a car. I guess the conceptual difference is that, rather than serve as the owner's manual for *one* EV, this is an owner's manual for *any* EV.

Because electric transportation is a spectrum of topics, and each of these topics is relevant at different stages of the electric-car-ownership experience, you can think of this book as a road map or guidebook to several of the considerations facing electric-car purchase and operation. Some of the topics, for example, are most relevant when considering an EV for the first time. On the other hand, some are more relevant after having owned an EV for a few months. Others still are of interest when thinking about your *next* EV.

This book's chapters are organized to first address common questions and concerns about EV ownership and then to move to more in-depth considerations. For example, Chapter 3 addresses questions of EV safety; everyone wants a safe vehicle — whether it's powered by batteries, hydrogen, or a flux capacitor — and Chapter 3 talks about just how safe most of today's electric vehicles are.

Still, it's not critical that you read the book from cover to cover in the chapter order presented. (I do, however, think you'll enjoy experiencing the book in its entirety.) You suffer no penalty for heading directly to concepts that interest you the most, although there's a chance you might need to jump to earlier chapters to review basic ideas or gain foundational context. I guess the penalty may be a cognitive one.

Anyway, web addresses, which are cited frequently as sources, appear in their own distinctive font. If you're reading a digital version of this book on a device connected to the Internet, you can click a web address to visit that website, like this: www.dummies.com.

Foolish Assumptions

For one existential reason above all others, I'd argue that the *idea* of electric transportation is relevant to nearly every human on the planet today — especially to those humans who are either too young to research electric vehicles or who may not have the financial wherewithal to purchase a personal vehicle.

That said, I'm working on the assumption that at least 95 percent of readers drawn to this title are now driving some kind of internal combustion engine (ICE) vehicle and are considering purchasing an EV as their next car, which I also assume will be sometime in the next three to six months.

A subset within that group of 95 percent may not be interested in an EV anytime soon, but may instead be looking for a single source of information about electric vehicles that will satisfy their curiosity without having to spend hours sifting through a trove of articles. Rest assured, I've done the research so that you don't have to. If this describes you, this book should do a very good job of addressing questions like these: "What's all this EV fuss about, anyway?" "Should I install upgraded electrical service in the garage of the house I rent out on Airbnb?" "My pension plan has a stake in three EV makers — is that gonna be good or bad for my retirement plans?"

For the other 5 percent: Why are you reading? I'd be curious to know, so feel free to drop me a line. My personal email is: hmsbrian@gmail.com.

Icons Used in This Book

As you make your way through this book (if that's how you're reading it), you see the following icons in the margins:

The Tip icon marks bits of information you will find particularly helpful. When you're skimming the book, these tips should pop out to give you a quick grasp of the topic.

Remember icons mark information that is important to keep in mind. Some of them review topics from earlier in the book that are relevant to the information being presented.

The Technical Stuff icon marks information of a technical nature that is more important to someone working in the field and who might need a bit more depth.

The Warning icon points out bits of information you can use to avoid problems you might encounter.

Beyond the Book

Because electric transportation is an evolving and complex field, there's no single source or best place to go for more information. Throughout this book, I leave you sources and suggestions for further reading and research. For example, I don't know exactly what cars will be offered by which manufacturers three years from now, or even which manufacturers will still be in business by that time, but I can point you to a few places where you will likely find the latest-and-greatest info — websites are easier to update than giant manufacturing outfits.

In addition to the introduction you're reading right now, this book comes with a free, access-anywhere Cheat Sheet containing information worth remembering about the joys of owing an electric car. To get this Cheat Sheet, simply go to `www.dummies.com` and type **Electric Cars For Dummies Cheat Sheet** in the Search box.

And in Chapter 13, I mention a specific website, which is also mentioned here — `www.brianculp.com` — and I'll use that space to keep you apprised of ongoing developments. The challenge of a book whose subject is a rapidly evolving, and indeed revolutionary, is that mentions of a specific make or model or incentive or price point can become dated by the time the pages have been glued together.

Where possible, I keep the subject matter focused on topics that should prove most relevant to EV buyers today as well as five years from now. For example, people buying today will want to know about range and battery life and how and where to charge. People buying in 2025 will want to have the answer to the same questions.

For everything else — the things that do change rapidly— there's the website.

Where to Go from Here

Get started reading Chapters 1 through 3 to help answer some of the most common concerns that people have when considering an electric car: what the main advantages are over internal combustion vehicles, how things like regen braking and over-the-air updates will keep the car performing better over time, and why an EV is likely the safest vehicle choice you can make.

Chapters 4 through 6 dive into the nitty-gritty details about an electric-vehicle purchase: what to consider when comparison shopping, weighing a battery-only vehicle against a plug-in hybrid electric car, insurance options, and even the age-old buying-versus-leasing dilemma.

Chapters 7 through 9 are where you go after that new EV is in your garage, or at least after you're seriously considering parking one there in the first place. Chapter 7 in particular covers everything a new EV owner needs to know about charging — how, where, and how fast. Chapter 8 provides advice about prolonging the battery pack, and Chapter 9 dispenses some best practices for keeping the EV in top form over the life of ownership.

Chapters 10 and 11 were a joy to write, because they deal mostly with the future. I talk there about what might be on the horizon when it comes to battery technology. I talk about that future day when none of us will have to drive a car any more than we have to drive an elevator. (Did you know that elevator operators used to be a thing? They were.) Most of the technologies presented in these two chapters likely won't come to pass, or at least not in their current iterations, but it doesn't make the speculation any less fun. (And also informative: one of the main takeaways from Chapters 10 and 11 is that it probably doesn't make much sense to wait for the technology that's "just around the corner," because the future tends to have plans different from ours.)

Finally, I delve into the topics of electric trucking and electric bicycles — electric bikes possibly proving to be humanity's most important electric vehicle yet.

We'll finish with some Parts of Ten that identify companies and countries that are leading the transformation (and name names about which are lagging), and even leave room for some EV predictions, most of which we can look back and laugh about come 2032.

Enjoy.

1 Getting In On the Popularity of Electric Cars

IN THIS PART . . .

Get your basic questions answered

Explore the technology behind EV innovation

Work with over-the-air updates

Determine exactly how safe electric vehicles are

» Benefitting from tech disruption

» Learning the lingo

» Seeing the key differences in owning an electric versus a gas-powered car

» Realizing that gas-powered cars won't be around forever

» Answering your top questions about electric cars

Chapter 1
So You're Thinking About Buying an Electric Car

With the aid of about 50 years of hindsight, I can unequivocally make this statement:

Life-changing moments rarely seem so at the time.

For me, one of the biggest events in my adult life went like this: In the summer of 2017, I went out for sushi and it changed my life.

Was the sushi really that good? No. But the *drive* to the sushi spot took place in a Tesla Model X — shout-out to Jonathan Guy and Ian Martinez for setting things up — and the experience lit my brain on fire.

Roughly one year after I took that first ride in an electric vehicle (EV, for short), subsequent events have led to some remarkable personal and professional transformations. There is a straight, bright line between that ride and a change of employers for my wife and me — we both now work for electric-only vehicle manufacturers. That ride also led to the purchase of — not one, but — *three* electric vehicles (with pending reservations for others) and the absolute certainty that we'll *never* own an internal combustion vehicle again, mostly for reasons that really have nothing to do with the environment, although we're certainly grateful about that added benefit of electric transportation. More about all that later.

That one night out eventually caused our family to leave our roots in Kansas City for the EV opportunities in San Francisco. (Yes, we miss the family and the barbecue, but we now spend weekends hiking among redwoods, taking our dog for walks on the beach, and taking advantage of our membership at a Sonoma winery — let's call it a wash.) That drive in that electric car led me to explore topics I'd *never* been interested in. It sparked a curiosity about electrical engineering and energy generation and rare earth metals and artificial intelligence and neural networks and even tangential subjects like city design and walkable communities and . . .

I'm getting ahead of myself.

As you can see, I get excited when I think about the transformative impact of electric vehicles, on both a personal scale and the global scale. And, because writing is thinking, I'm sure I'll have a hard time containing that excitement as I put the story of electric transportation on the page. I promise I'll do the best I can, given the time and space constraints of a *For Dummies* book. (Note that such constraints are entirely appropriate; no single resource can cover all issues related to EVs. There are single paragraphs in this book that are the focus of multiple years of university study. What's more, any resource that even tried to be a compendium on electric transportation would be hopelessly out of date by the time it was ready for publication.)

Oh, and I almost forgot: That sushi dinner led to this book.

In the pages that follow, I have the chance to share some of the experiences I've had with electric cars over the past four years, experience being the best teacher of what EV ownership is like. Additionally, I share some of the "inside baseball" details I've gathered by way of thousands of conversations with really, really smart engineers while working for three years at one particular electric car manufacturer (and talking about another manufacturer most evenings with a spouse who is also a really, really smart engineer).

To say that I'm excited about the pages that follow would be an understatement of gigawatt-hour proportions.

So, will picking up this book be as transformative for you as that first ride in an electric vehicle was for me in the summer of 2017?

Only one way to find out.

Why Buy an Electric Car?

Hmm . . . the answer to the question of why you should buy an electric car would fill many, many pages — 384 of them, to be exact — which science has proven is the *perfect* number of pages for a book on the topic of electric cars.

The short answer is that there are dozens of reasons it would make sense for your next car to run on electricity (or, in the case of a plug-in hybrid, to run on electricity for at least a significant chunk of its daily travels). You could buy an EV because you want to reduce the costs to fuel the car, leaving you more spending power every month. You might want only to visit a convenience store for the coffee, and not for the experience of handling a gas pump in either the freezing cold or the rotisserie-chicken heat. You might even be mindful of the geopolitical implications of buying gasoline — where it comes from and what kinds of activities your purchase might be helping to fund — concluding, *à la* the movie *WarGames* (kids, it's a 1980s film about the futility of nuclear war), that the only way to win is not to play. You might be motivated to spend less time in the shop, waiting for repairs and maintenance. Or, it might simply be time to replace an old beater and you want to take advantage of the tax incentives. You might even be one of those serious-car-enthusiast types who wants bragging rights over all your muscle car friends when talking about 0-to-60 times. (See Figure 1-1.)

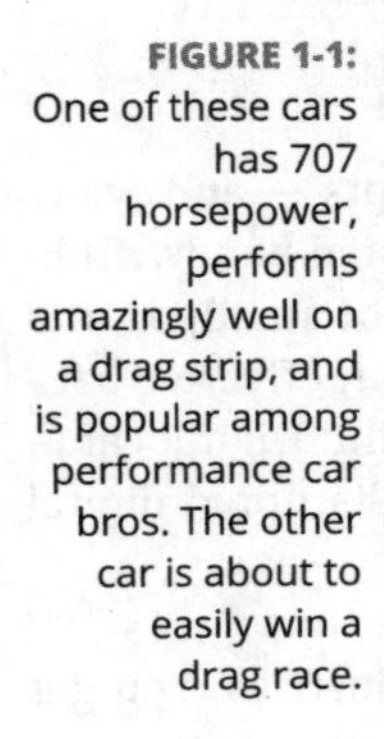

FIGURE 1-1: One of these cars has 707 horsepower, performs amazingly well on a drag strip, and is popular among performance car bros. The other car is about to easily win a drag race.

And, of course, you might be sitting across the dinner table one night from your 12-year-old child — maybe you'll take your son or daughter out for sushi — when you'll suddenly ask yourself: What kind of planet will this child be living in when they're my age? I may not be able to change the world all by myself, but if I can play some small part in humanity's job of reducing its collective carbon footprint, don't I owe it to the child to make that effort? "I'm sorry it's 140 degrees outside and that most of the planet's crops have failed, son, but you should see my 401k!"

So, there are lots of reasons to drive an EV, and you'll get a chance to explore each one of them in the chapters ahead. But, for now, you don't even have to take my word for it — no less an automotive authority than the American Lung Association also recommends that humans with lungs, or at least humans who have other humans they care about who rely on lungs, make the switch to an EV because ". . . EVs are pleasant to drive. EVs generally have a lower center of gravity, which offers better handling, comfort, and responsiveness. The electric engine provides smooth acceleration and deceleration, and a quiet ride, which all leads to a better driving experience."

So there. Cancel your subscriptions to *Motor Trend* and *Car and Driver* magazines, you serious-car-enthusiast types. And stop browsing articles on Electrek and Greencarreports, you curious-consumer-who's-just-researching-some-options types. If the American Lung Association says to drive an EV, then let's get with the program here, people.

Because, at the end of the day, all these reasons to buy an electric car that the American Lung Association cites are reasons because of one simple fact:

It's a better technology.

The Disruption of Better Tech

Throughout the course of human history, better technology disrupts — and eventually replaces — the older technology. The printing press disrupted handwritten books and papers. Vaccinees disrupted millions of people dying of smallpox and measles. Gunpowder disrupted calvary charges on horseback as an effective battleground tactic except in *Lord of the Rings* movies. Speaking of film: Digital cameras disrupted the film processing business, and smartphones disrupted digital cameras.

And, cars disrupted both the horse-and-carriage and farrier industries. You get the idea.

REMEMBER

Even calling things disruptive is, um, disruptive. The term, at least as applied to existing technology and value networks, is credited to the Harvard Business School professor Clayton Christensen and his work on the theory of disruptive innovation (published in *Harvard Business Review* in 1995), and it stems from his work on the disk drive industry. Hey, remember floppy disks? Remember when they were disrupted the zip drive made by a company called Iomega? Yeah, me, neither. I mean I'm not *that* old, except that I am. (Curious about Mr. Christensen? You can check out the original article at `https://hbr.org/2015/12/what-is-disruptive-innovation`.)

So, whether you decide to purchase an electric vehicle within a few weeks or within a few years after reading this book, or perhaps you decide not to buy a car at all, in favor of ride-hailing and/or per-use rentals, you and the rest of humanity will be either driving or riding in electric vehicles relatively soon. (If it hasn't already happened, that is. I trust that some of my audience will have just purchased their first EV and will have come here for a few pointers — and then will stay for the sparkling prose throughout.)

Why will humanity choose electric vehicles (EVs) over internal combustion engines (ICEs)? Is it because all of humanity will rally together to save the planet, collectively deciding to do everything they can to live more sustainably?

As much as I'd like that to be the case, I have been sentient these past ten years or so (sometimes regrettably so) and have grown rather pessimistic about our collective human ability to cooperate on issues whose benefits would be enjoyed by future generations.

Fortunately, the benefits of driving electric will accrue to even the most self-centered among us. The rate of electric adoption will increase between the years 2022–2030, for two main reasons:

- Electrics are cheaper to own and operate.
- Electrics are *way* more fun to drive.

So the real answer to the question of why choose EVs over ICEs is the same as it is for why choose the iPhone over Blackberry or why choose Netflix over Blockbuster.

Because it's better tech, resulting in a better consumer experience.

In other words, consider the inevitability of electric transportation in terms of the inevitability of smartphones (over landlines), broadband Internet (over dial-up), streaming music (over compact discs), or any of several dozen other technologies that have been replaced over the years by better, cheaper, faster, or more convenient alternatives.

REMEMBER

What is technology, anyway? People use the term often in oral and written communication, but they seldom stop to consider its implication. Briefly defined, *technology* is applied science.

For example, science is learning, through trial-and-error (while trying to control for only one variable at a time), that certain metals react to heat such that they can be shaped or even melted, or that other metals are good at storing negatively charged ions.

Technology is engineering something that's based on that science, like shaping molten steel into a plow in order to improve agriculture or using lithium in the manufacturing of a battery capable of being recharged, which in turn improves mobility.

Or, for the Adam Smith *laissez-faire* capitalists in the audience, you can answer the question of "why EVs?" using a purely market-driven framework. Companies like Tesla, Rivian, Nio, XPeng, and Lucid have recently burst onto the scene (at least in terms of public awareness; each has been *around* for many years) while giant, established automakers like Ford, Chevy, Volkswagen, and Hyundai have announced either majority-electric or all-electric futures. (Figure 1-2 shows Adam Smith in all his 18th Century glory.)

Hulton Archive/Getty Images

FIGURE 1-2: "Rational self-interest says you will drive an EV," is not a *direct* quote from Adam Smith, but at the same time it is *so* Adam Smith.

They're not making these multi-hundred-billion-dollar commitments simply to garner headlines, or to flex about their green street cred. They're doing it because they think electric cars are what consumers will *demand* over the long term. As with any other company, these auto manufacturers' sole purpose is ultimately to meet consumers where they are and eventually make a profit by meeting that demand. I do acknowledge that this is a combination of art and science, which involves varying degrees of market research — looking backward to spot patterns — and market forecasting — either using the patterns identified in market research to project future needs, or just having a strong conviction about something you *think* people will love. There was no market study that would have supported Howard Shultz's opening a specialty coffee shop selling espressos and lattes for example, but here we are, with a Starbucks on every corner.

And speaking of disruption: I use a couple of acronyms in the earlier section on disruption — namely, EV and ICE — which I assume are common knowledge to those curious about electric transportation. That said, that won't be the last time you see an acronym in this book. So I'm going to disrupt the train of thinking about EVs that just connected farriers, Netflix, Adam Smith, and, by inference, Starbucks Coffee and lay out a few of the acronyms that will facilitate a better understanding of the topic at hand. All so that you won't be disrupted every time you come across an acronym.

Plus, it all connects with Homer Simpson because — why not? (I warned you I can be enthusiastic about this stuff.)

Acronyms and Abbreviations That Make You Sound Smrt

As a technical writer, I tend to detest acronyms, mostly because of the assumptions baked into their use.

That said, there are a couple of abbreviations that I end up using throughout this book — abbreviations I expect you to know as a matter of course. ICE, for example, which I use in the previous section, is the acronym for *i*nternal *c*ombustion *e*ngine, and this is probably the only time I'll pause to point that out. I will assume that 99 percent of that audience for a book about cars has encountered the acronym ICE and wasn't thinking of frozen water when seeing it here.

TIP

In case you're wondering, the section heading is borrowed from a single Homer Simpson line, which he sings after being accepted to college. Google the phrase *Homer Simpson I am smart* to feel at one with all the cool kids who picked up on the reference immediately.

Besides setting the foundation for what follows, another reason I'm pausing for this brief reference section is that I assume this won't be the only source of electric vehicle information you ever consult and that many of those sources will use these same abbreviations. Knowing these acronyms helps you better absorb these other sources of information. You're welcome.

The following list, then, covers many of the more common acronyms you'll find when learning about EVs:

- **EV:** Electric vehicle (*see also* BEV, a couple of lines down)
- **ICE:** Internal combustion vehicle
- **BEV:** Battery electric vehicle — a 100 percent electric vehicle
- **PHEV:** Plug-in hybrid vehicle — a vehicle whose drivetrain can be powered by a battery as well as by a gasoline engine
- **HEV:** Hybrid electric vehicle — an ICE-first vehicle that also has the capability of running off a limited battery. Hybrids don't have a plug. They generate all the electricity stored in the battery from the power generated by the combustion engine, and through regenerative braking

Figure 1-3 shows in graphical form how these vehicles differ.

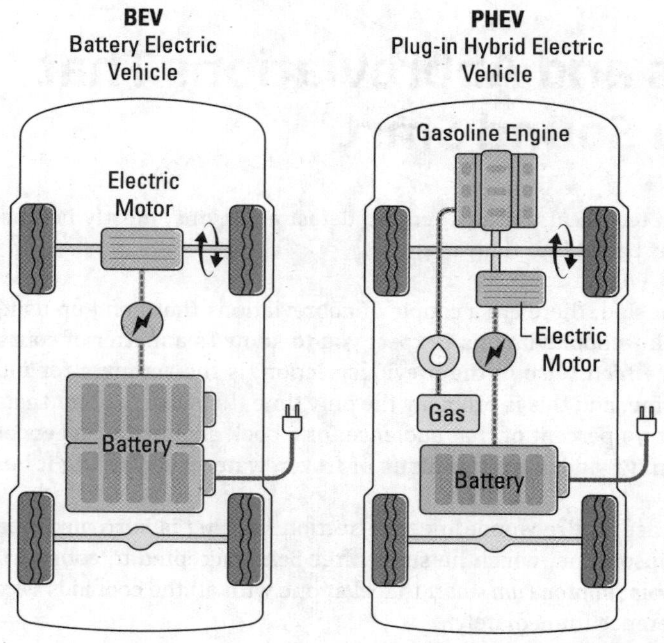

FIGURE 1-3: EV means no ICE. PHEV means EV *and* ICE.

Source: VectorMine / Adobe Stock

REMEMBER

Most of the book you're holding deals with battery-only electric vehicles. When you hear *EV*, the generally accepted definition is that it's referring to a battery-only automobile or truck.

Now that you have confidence that you won't be tripped up by the acronyms in a phrase like "EV versus ICE," turn your attention back to the main business of this introductory chapter and indeed the entire book. Let's examine some of the differences between the two kinds of vehicles.

Drawing the Line between Electric and Gas-Powered Vehicles

Before I even try to describe in words how EVs and ICEs differ, just know that anything I write here ultimately doesn't matter nearly as much as the tactile experience of getting behind the wheel of an electric car and driving it for yourself.

I say that with the confidence that comes from knowing in my bones that one particular aspect of human behavior is true, even though I have no study I can direct you to that will back up my claim, nor am I even aware that this particular aspect has an official name.

So I'm giving it an unofficial name: It's Brian's *Book of Boba Fett* Law. (It's a corollary to the Law Governing All Consumer Products and Mass Entertainment, which is yet another law that exists entirely in my head.)

Brian's BOBF Law goes like this:

> **People don't know what they want until you show it to them.**

If you're thinking right now that Steve Jobs uttered something similar during some speech given somewhere and that I just stole that line and repackaged it with a pop culture reference, know that you *could* be right — about the stolen line, that is — but there doesn't seem to be an authoritative answer about whether Jobs actually said this, or whether the quote is apocryphal.

TIP

In the same purported speech from Steve Jobs, he also purportedly quoted Henry Ford, who purportedly expressed something similar, saying "If I'd have asked people what they wanted, they'd have said faster horses." Except Ford *definitely* never uttered that phrase, despite being famous for it today. So thank you, Internet — I guess. In any event, researching all this took me two *Inception* levels down into the Internet's dreamworld of quote misattribution. Should you explore that second level, you'll emerge assuming that anything in an email signature attributed to Einstein, Gandhi, Ben Franklin, Steve Jobs, or a dozen other leaders and thinkers has been entirely made up. There is no fact-checking embedded in a copy/paste operation.

For the purposes of discussing electric cars, it doesn't matter.

Brian's *Book of Boba Fett* Law posits that audiences don't want to see a series about Boba Fett. Rather, they want to *feel* the same sensations they felt when experiencing Boba Fett onscreen for the first time in 1982. They want that sense of awe, excitement, and possibility.

Driving an electric car for the first time is like seeing Boba Fett in *The Empire Strikes Back*. (See Figure 1-4.) After experiencing an EV, subsequent drives in ICE vehicles feel like seeing Boba Fett in *The Book of Boba Fett*.

Anna Lurye / Adobe Stock

FIGURE 1-4: Because an actual pic of Boba Fett is a sure fire way to get sued by Disney, we're using this photo instead and asking you to conjure an image of old Boba in your head. The imaginative result is the *Star Wars* equivalent of thinking that people want faster horses.

Beyond that, I'm going to spend the next 300+ pages exploring the differences between an electric car versus a gas-powered car. If you read this book from start to finish, you'll have a lot of information that you didn't have when you began. What will be almost impossible for me to replicate, however, is the sense of awe, excitement, and possibility you'll feel when you get behind the wheel of an EV for the first time.

When you drive one, you'll get it. You'll understand that this is the thing you wanted — it's just that no one has showed it to you until now.

And, of course, none of this is new in terms of humans and their technology. People don't know what's on technology's next page, yet they often like the words written once they get there. For example, people didn't know they wanted a refrigerator to replace their iceboxes. They didn't know they wanted air conditioners or backup cameras or seat heaters or Bluetooth connectivity in their cars . . . until the first time someone installed one. And I was perfectly happy with the occasional coffee from a convenience store until I walked into a Seattle coffeehouse named after a character in *Moby Dick*. (True story: The first time I visited a Starbucks, it was the original Seattle Public Market location.)

So, as I said, I have another 300+ pages to unearth all the glorious details about the differences between EVs and ICE.

I'll close the thought, then, by referring to something Steve Jobs definitely *did* say, when talking about the Macintosh in 1985:

> "We built it for ourselves. We were the group of people who were going to judge whether it was great or not. We weren't going to go out and do market research. We just wanted to build the best thing we could build."

(You can find those lines in a write-up in *Inc.* magazine about Jobs' approach to keeping the customer satisfied (`www.inc.com/jason-aten/this-was-steve-jobs-most-controversial-legacy-it-was-also-his-most-brilliant.html`), although it's secondhand. The actual quote comes from, believe it or not, a *Playboy* magazine interview.)

If *you* built a car — any car — you'd likely build a car that was quicker than the one you have now. You'd (likely) build one that's quieter — one that's safer if it gets into an accident. You'd build a machine that emits zero greenhouse gases and is also much cheaper to maintain and much cheaper to fuel and can even be refueled from the comfort of your own home. You'd want a car you can upgrade with a new operating system, like your phone — a car that can improve characteristics like range and speed with a few lines of code. And you might even build one that can drive itself around so that you can use drive time to catch up on work.

In short, you'd build the best thing you can build. And you'd start that building process with an electric motor and a battery. And then, after you'd built that new, great thing, here's what you'd tell others about it. Here's what you'd answer when people asked you why they should drive your automobile and not the one they already have:

» **EVs are better to drive:** That battery-and-electric-motor contraption in your driveway is an absolute blast to drive. And, if you're parting with your cash in exchange for a driving machine, you might as well get the best driving experience your money will buy. They're quiet. They're quick. They tend to handle exceptionally well, thanks to their low center of gravity.

  EVs provide the kind of driving performance I'm talking about in the places where it's most keenly felt by most people: when pulling away from a stoplight (the Camaro, Mustang, and Challenger don't stand a chance against most EVs) or merging into highway traffic at speed or quickly passing in the highway's fast lane.

Plus, there's a nonzero chance that, whenever you grow tired of driving your EV, you can let the EV do the driving for you instead. (Oh, you also put an array of cameras and radar into the car you built, and you know how they always say that there's more computing power in your phone today than was used to put a man on the moon? Yeah. You've leveraged that and now your car does enough floating-point calculations to run an entire space program. Or, at the very least, to watch Netflix.)

- **EVs are better for your wallet:** EVs are much less expensive to operate on a daily basis. First, as extraordinarily cheap as gasoline is — it's about $6 per gallon in California as I write this section, and yes, that's extraordinarily cheap — gas is no match for electricity when it comes to cost. Electric rates change very infrequently, and when they do, it's in very small percentages (unless you live in Texas during an ice storm, I suppose). When we owned our EV in Kansas, we paid 13 cents per kilowatt hour for electricity. In California, we pay 16 cents during off-peak hours.

  In terms of impact to the pocketbook, it now costs around $100 to fill a typical fuel tank in an ICE vehicle (these include cars and pickups). In the typical EV, it costs about $10 to charge the battery. (A Toyota Tundra, with a 32-gallon tank, costs $160 to fill up when gas prices are $4.99 per gallon. Is the Tundra an outlier? Yes, but at California's current average price for gas, Tundra owners pay over $200 to fill 'er up. And I can report this as gospel because my cousin owns a Tundra. Meanwhile, a 16-gallon tank at California's average price as this goes to print comes in at $104.

  I cover all the exact math I use to get to these numbers in Chapter 4 (and elsewhere). Oh, and I haven't even touched on the maintenance costs, which I do in Chapter 9.

- **EVs are easier to fuel:** Your new electric vehicle invention is not only cheaper to fuel — it's also much more convenient for most households. It doesn't, for example, require you to stand and hold a gasoline nozzle in a rainstorm, or in 110-degree heat while petroleum fumes turn your brain into nacho cheese.

  For the most part, you'll treat your EV the way you treat your Bluetooth headphones or your laptop: You'll plug it in while you're not using it, which for the most part is at night. You'll then wake to a laptop battery that can power through a long day of meetings and a car battery that can power through a long commute to and from the office.

- **EVs are better for the planet:** Whether or not you care about the relatively insignificant impact of a single car on the health of the environment, I promise you that your kids and grandkids do very much care. As do the kids and grandkids of the neighbors on your street. And the kids and grandkids of people you've never met on the other side of the planet.

No argument or even civil discussion about society, the economy, or, really, any topic under the sun will even matter if the sun, with the aid of greenhouse gases, has made the planet uninhabitable for humans.

As mentioned earlier, EVs don't produce tailpipe emissions. They have a much smaller carbon impact over the life of the vehicle than their ICE counterparts.

We humans may be unable to do anything about the changes to planetary ecosystems that science is telling us is heading our way, but if your actions can be a very small step in the direction of mitigating those consequences . . . isn't that a step worth taking? Don't you owe that to your kids?

Yeah, but Gas Cars Aren't Going Anywhere for a While — Wait, What?

One somewhat reasonable response to all this talk about electric cars — what can sound like obnoxious, techno-utopian gushing in the ears of non-EV converts — is to dismiss it as much ado about nothing. We were all gonna have flying cars by now, and robots to do all our housework, and colonies on the moon, and blah, blah.

Granted, few readers who pick up this book will make this argument, but the argument itself goes something like this:

> *Yeah, I drive down the road and see just a couple of EVs here and there. And, fine — they might be quicker and such, but cars are big purchases and gas stations are still everywhere and EVs have been available for a good eight years and I still don't see that many on the road, so we're only talking about 1 or 2 percentage points of adoption per year, which means that this is mostly a bunch of greenwashing hype. People such as the idiots who write books when they should be making TikTok videos like a normal person does are saying that EV adoption will be at the 50 percent mark by 2030, but 1 or 2 percentage points per year will only get us to, what? about 15 percent by 2030 — 20 percent, tops. Thanks, but I'll keep my SUV for now.*

Again, it's not an unreasonable argument as far as strawman arguments go. (And, of course, setting up a strawman is at times a rhetorical necessity; there of course isn't one person making the claim I just laid out. Rather, it's an amalgam of explicit and implicit criticisms of EV enthusiasts like me.)

The counterpoint to that general line of thinking is shown in Figure 1-5:

Or in Figure 1-6.

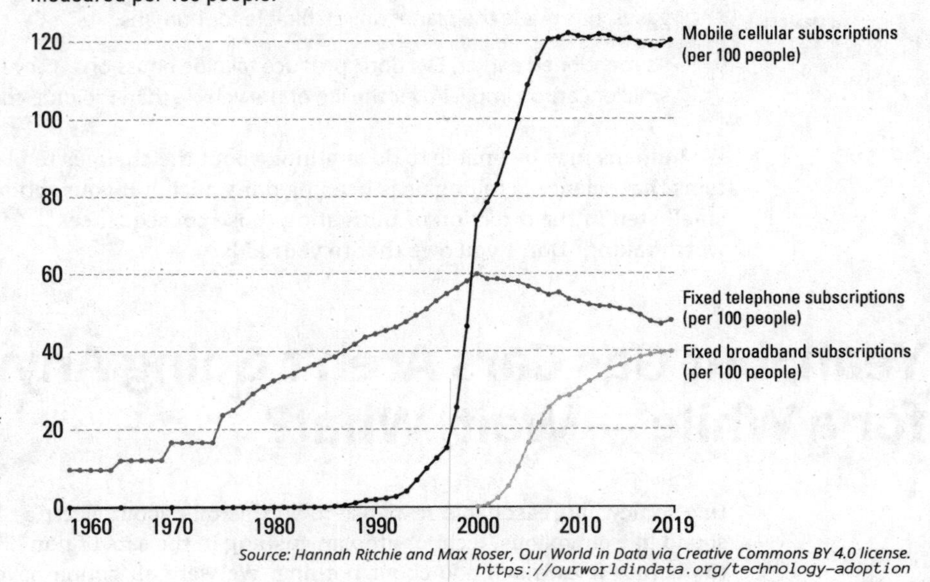

FIGURE 1-5: The S curve of better tech — mobile phone adoption in the UK.

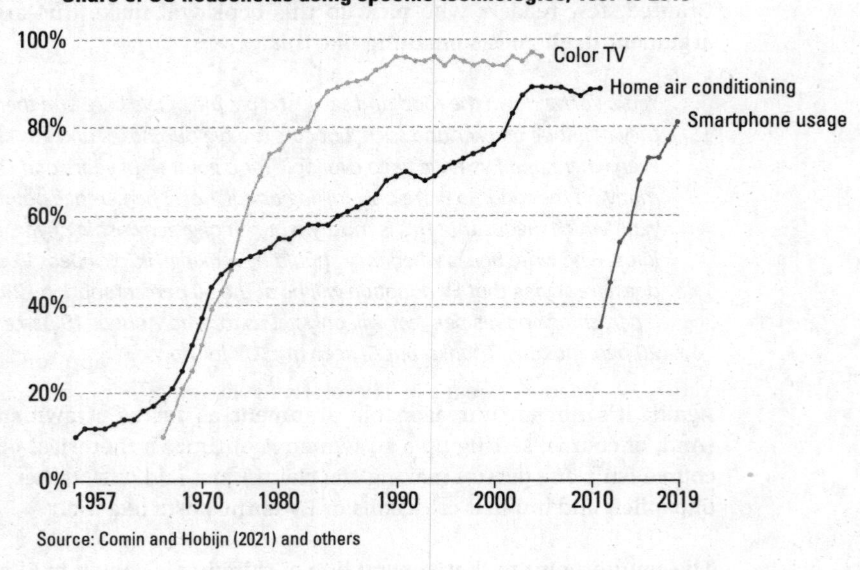

FIGURE 1-6: The S curve of better tech — refrigerators, color tv.

Or even in Figure 1-7.

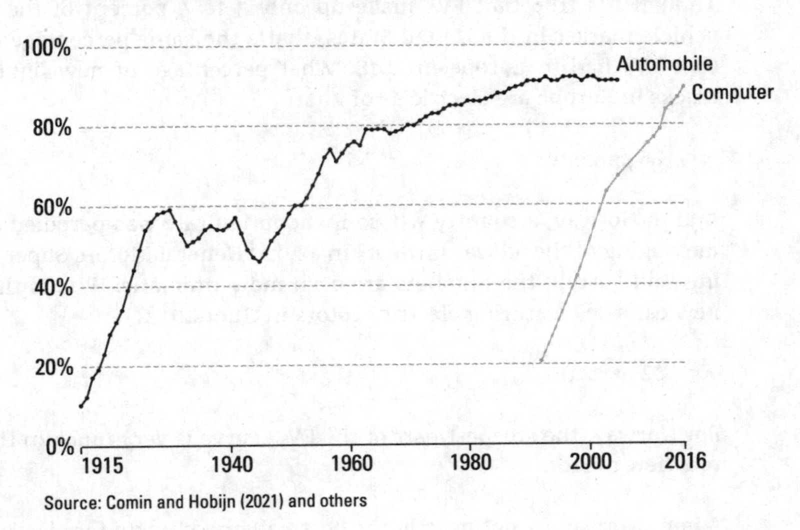

FIGURE 1-7: The S curve of better tech — autos and computers.

And note that I've placed this last chart alongside a backdrop of automobile adoption, itself a very disruptive technology, especially during the decade between 1915 and 1925.

The point is that disruptive technologies always follow an S curve of adoption. It's an immutable law of consumer technology: If it doesn't follow an S curve, it's not truly disruptive.

And the velocity of that curve has become even more pronounced over the past 50 years, for everything from credit cards to personal computers to smartphones to the apps on those smartphones. About 10 years ago, before it became a mechanism to depress young girls and enrage young men (and not a few old ones, either) for fun and profit, Facebook was a Hot or Not app for elite college campuses. And had you heard of TikTok five years ago?

Me, neither.

Where gas cars are going in Europe (and especially Norway)

Though it's true that EVs make up only 3 to 4 percent of the new light-duty vehicle market in the United States, that's the same percentage of market share that EVs had in Europe in 2018. What percentage of new light-duty cars and trucks in Europe are electric as of 2021?

26 percent

And in Norway, a country whose EV adoption rate was parodied (I guess? — the messaging of the ad was unclear) in a 2020 General Motors Super Bowl ad featuring Will Farrell, the numbers are even more dramatic. What's the percentage of new car sales featuring electric motors in Q1 of 2022?

82 percent

For Norway, the steepest part of the EV S curve is very much in the metaphorical rearview mirror.

Again, Norway — not exactly the best <ahem> climate for electric vehicle adoption, what with the bone-chilling winters and how they generally impact battery life. What Norway does have going for it is this:

- **Gasoline prices:** Though still lower than a truly free market dynamics might dictate, gas prices *are* more in line with the true cost of extracting, refining, storing, and transporting the stuff.
- **EV supply:** There's at least enough supply to meet the car-buying demands of a relatively small population, as far as countries go.

So, does that mean that the rate in the US will grow from 3 to 4 percent in 2022 to 26 percent by 2025? Can we hope to emulate Norway by 2030?

Maybe.

For one thing, Europe had some things going for it in the EV space that haven't yet happened in the US. The catalyst was perhaps that Europe began enacting stricter emissions standards in 2019, and this incentivized manufacturers to bring more EVs to market in Europe.

However, consumers quickly learned that the experience of driving electric was superior to driving ICE. (See also pretty much everything I've written so far in this chapter.)

As a result, demand exceeded linear forecasts that were modeled purely on compliance with tighter emission standards. In other words, the models predicted how many consumers would modify a purchase decision in order to take advantage of the tax savings meant to incentivize voluntary compliance. (See Figure 1-8.)

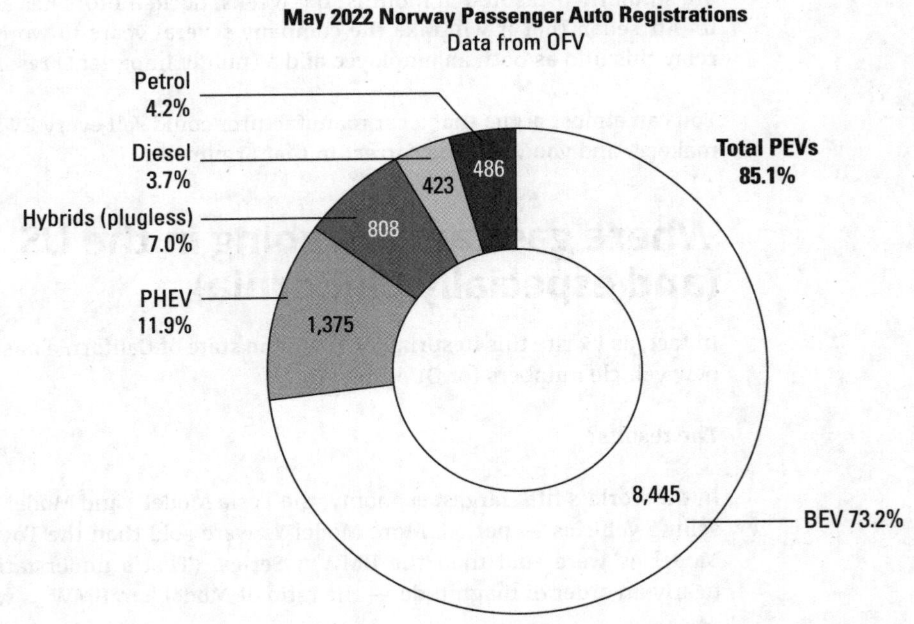

FIGURE 1-8: As it is in Oslo, so it is for the world entire — at least one can hope.

The US has none of these same vehicle emissions standards in place, however. As a result, most major auto manufacturers have prioritized Europe for EV availability. As a result of *that*, the US simply has a limited number of EVs available for purchase. As a result of *that*, Tesla keeps gaining market share in the US. At exactly the moment the US's demand curve has started to hit the steep part of the S, Tesla has been effectively the only manufacturer there to answer the call.

And make no mistake: Demand in the US market has almost *certainly* reached the tipping point where the S curve begins the steepest point of ascent. Tesla alone delivered nearly 1 million EVs in 2021, and in 2022 customers are routinely having to wait almost a year to have a new Tesla delivered, despite steadily rising prices. When Elon Musk isn't trolling the elected officials who support the tax credits that were essential to helping the company establish its footing, he's even hinted at the possibility of *stopping* reservations to give the company time to work through the backlog of orders.

Wait times for the Hyundai Ioniq, Rivian R1T, Ford F-150 Lightning, and others are similarly measured in months, not weeks. Lucid Motors has an order book for its Air sedan that it will take the company several years to work through, and I relay this info as both an employee and a (mildly impatient) reservation holder.

You can almost argue that a car manufacturer could sell every EV it could possibly make. . . and you would be correct in that argument.

Where gas cars are going in the US (and especially California)

In fact, as I write this in spring of 2022, the state of California has just released its new vehicle numbers for Q1 of the year.

The results?

In the world's fifth largest economy, the Tesla Model 3 and Model Y were two top-selling vehicles — period. More Model Ys were sold than the Toyota RAV4. More Model 3s were sold than the BMW 3 Series. (That's understating the case by nearly an order of magnitude — the ratio of Model 3 to BMW 3s was over 8 to 1.)

Take that, Norway.

A light-duty vehicle is a legal classification that specifies a maximum gross vehicle weight rating (GVWR) of less than 8,500 lbs. in the US. It's a term mostly used to describe trucks, but passenger cars are also light-duty vehicles.

California also has the highest gasoline prices in the US, which is certainly a contributing factor to the rate of EV adoption. (See Figure 1-9.) But, as I explore further, fueling with even the cheapest gallon of gas is way more expensive than fueling with electrons.

FIGURE 1-9: This sign is the subject of a lot of news headlines, to which many Californians reply, "What's a gallon of gas?"

AP Photo/Jeff Chiu

TAKE <u>THIS</u> ON, NORWAY

I'm not done gushing about Norway.

The reason why is because part of Norway's story of electric vehicle transformation involves the lead singer of the 80's synth-pop band A-Ha, and it's just way too cool of a story not to pass along. I wouldn't be doing my job if I didn't share this far and wide.

It goes like this: One day in 1990, A-Ha's a lead singer Morten Harket, who was also the head of the Norwegian environmental group named Bellona, got fed up with his country's policies towards the adoption of electric vehicles.

He then channeled his inner Forrest Gump and got behind the wheel of a converted electric Fiat Panda (see figure below) and started . . . just driving around Oslo.

With this car, Harket did indeed *Stay on These Roads*. (And if you get that reference, then you indeed are a true A-Ha fan and child of the 80s. Bravo, sir or madam. Bravo, spirit animal, you.)

But of course, that wasn't all. During the drive, Harket also refused to pay Oslo's notoriously high road tolls, parked illegally wherever he could, and promptly ignored every penalty notice issued to him, so much so that eventually the authorities impounded Harket's car and auctioned it off to cover the fines.

(continued)

(continued)

Except! There was only one buyer at the police auction for a converted electric Fiat Panda: Morten Harket.

And can you guess what a relatively wealthy (royalties from "Take on Me" could have funded a lot of these civil disobedience adventures, one assumes) pop singer does when he gets his electric Fiat out of impound? The answer is "Whatever the heck he wants," which in this case meant more driving, more illegal parking, and more ignoring of traffic fines.

Harket's act of protest attracted quite a bit of coverage — this was all pre social media, or course — and the point was made.

Soon, electric vehicles were exempted from road tolls, one of a large raft of incentives that have, over the years, helped make Norway the country with the world's highest per capita electric vehicle ownership. The government began offering incentives to buy and run electric cars as far back as 1990, first by introducing a temporary exemption from Norway's exorbitant vehicle purchase tax, which became permanent six years later.

Note: After initially disbanding in 2010, A-Ha has since reunited and is producing new work as of today. They have a new film and an album, both called *True North*, scheduled for release in October 2022.

So, thank you Norway for your leadership on electric vehicles. And thank you Morten Harket, for your leadership of Norway, for showing Norway's leaders that they could do both the right thing and the cool thing at the same time.

august.columbo/ Adobe Stock

Why predictions are unpredictable

As we get familiar with the idea of an S-curve and ponder how that almost certainly applies to electric vehicles, here's the thing to remember about market predictions: People — and by *people,* I mean people who get paid to make these sorts of predictions — tend to *underestimate* the adoption rates of transformative technologies.

Here's an example that I'm borrowing from a mind-splintering presentation I watched a few years ago, given by Tony Seba, the founder of an energy-and-transportation think tank called RethinkX (and no, I don't really know what a think tank does, either).

TIP

Here's the link to the specific presentation I'm referring to for those curious: `https://www.youtube.com/watch?v=2b3ttqYDwF0&t=200s`. Recorded in 2017, yet still relevant — if not downright prophetic — it's well worth an hour of your time, especially if you're into having your mind completely splintered by new ideas. (There are more recent versions; each is essentially the same presentation.) I thought about that presentation almost every day for literally two years after I first saw it.

Anyway, the market prediction in question here has to do with cellphones, and the expert predictors involved here are the American Telephone and Telegraph Company (AT&T) and the market research consulting group (is that too a think tank?) McKinsey & Company.

In 1985, AT&T hired McKinsey to forecast the market adoption rate of this new creation (in which AT&T had a passing interest), something called a "cellphone." McKinsey's task was to predict US cellphone adoption 15 years into the future, which would have taken them out to the year 2000.

McKinsey's prediction forecast of US cellphone adoption by Y2K?

> 900,000

Given that the population of the US is 325 million and given that it now seems as if every man, woman, and child over the age of 5 has a cellphone glued to their bodies 24/7, you probably know where this is going.

The actual US cellphone adoption when the calendar hit January 1 and the world's computers didn't shut down (in yet another bit of prediction that didn't age well)?

> 109 million

As Seba points out, McKinsey missed the projection by a factor of only 120x. So, yeah, even experts whose job it is to get things right (billions of dollars and countless man-years of labor can be mobilized based on these predictions) get things wrong — and wildly so.

That said, I also want to point out three things here, if only to give you a sense of what kind of author you're placing your trust in for the next 300 or so pages:

» **This AT&T/McKinsey study is just a single data point.** Predictions are a tough business. And there are plenty of examples of market predictions that have erred in the other direction. Nuclear fusion. Heck, nuclear power in general. The adoption of Betamax over VHS. The market adoption of hydrogen-powered cars. The abiding fashion styles of Von Dutch and Ed Hardy.

  Wherever practical and possible, I direct you to the original source of a claim, like the one I just mentioned about AT&T and McKinsey. "Trust but verify," an old Russian aphorism, is my approach to compiling this book. In the case of Seba's information about the potential cellphone market prediction, the claim checks out. I don't have the original market research at hand, of course, but this anecdote is verified in reporting by the *New York Times* and *The Economist* `www.economist.com/special-report/1999/10/07/cutting-the-cord`.

» **Experts are experts for a reason.** If you're looking for an author who puts "experts" in air quotes (or even printed air quotes) or tells you never to trust the mainstream media, you've come across the wrong guy. Experts can make mistakes, and making predictions is, as I've mentioned, extremely difficult. And, for the most part, the mainstream media works very hard to get the reporting right so that I don't have to exhaust myself fact-checking every darn word I ever come across.

Now then: As we hit the home stretch of this chapter, let's return in our minds to the cost curves presented earlier in this section, take a look at some interrelated factors in terms of EV adoption, and see where all this is leading us.

Let's begin with the cost curve of solar. The cost of solar panels has been in steady decline over the past decade, outpacing the predictions of the International Energy Agency (IEA). (See Figure 1-10.)

What's this mean?

It means that although gasoline is subject to wild price fluctuations in the present and future, just as it has been in the past (ask your parents about the OPEC crisis of the late 1970s), the price of electricity is extraordinarily stable, and is far less likely to be impacted by inflation than almost any other consumer good.

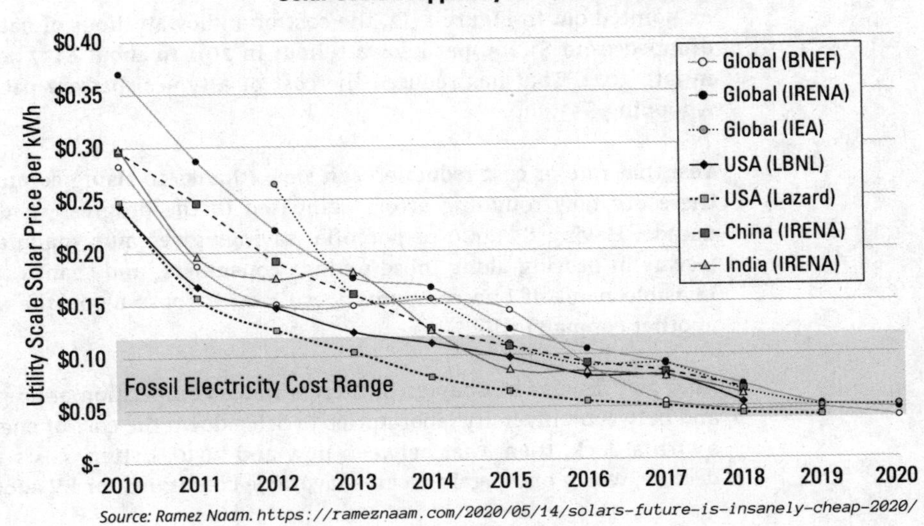

FIGURE 1-10: The S curve of lower solar costs.

Source: Ramez Naam. `https://rameznaam.com/2020/05/14/solars-future-is-insanely-cheap-2020/`

So EVs need electricity, and low, stable electricity costs are favorable to EV adoption. What else? What else do EVs rely on to move from place to place?

Oh, yeah. Batteries. What's that cost curve look like? Figure 1-11 has the answer.

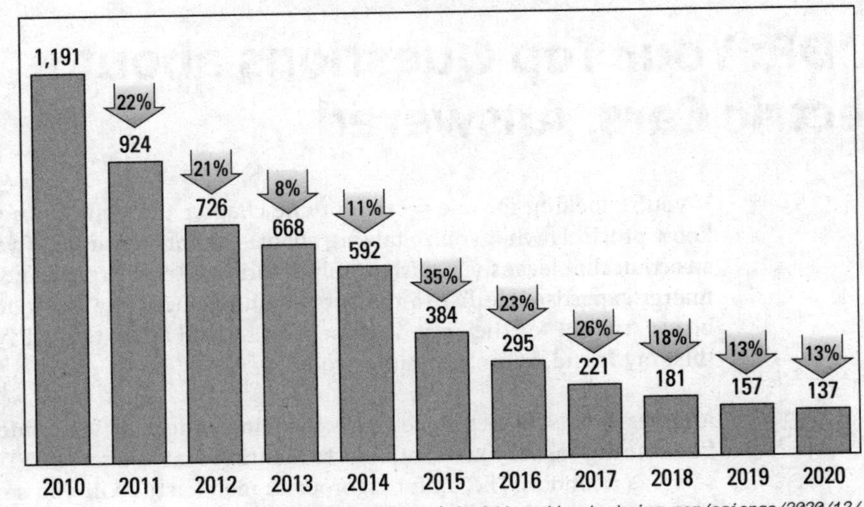

FIGURE 1-11: The S curve of lower costs — is all this starting to add up now?

Source: BloombergNEF/Arstechnica (`https://arstechnica.com/science/2020/12/battery-prices-have-fallen-88-percent-over-the-last-decade/`)

As pointed out in Figure 1-12, the cost of a kilowatt-hour of battery storage has dropped from $1,200 per kilowatt-hour in 2011 to about $137 per kilowatt-hour in late 2021. That has reduced the cost of a typical battery pack for a car by a whopping $53,000.

Yes, that rate of cost reduction can slow, thanks to rising commodity costs, but these are now rounding errors compared to the progress made over the past decade. Having $53,000 of potential savings gives auto manufacturers a lot of leeway in passing along those savings consumers, and thanks to Adam Smith's invisible hand, if one company doesn't do it (pass along the savings that is), another company will.

And, as I discuss in Chapter 8, there's fierce competition among manufacturers and between university laboratories to drive down the cost of energy storage. It's a virtual lock, then, that between now and 2030, battery costs will continue to decline, which once again creates favorable conditions for EV adoption.

When you start assembling the pieces of the electric vehicle puzzle, a clear picture emerges: ICE vehicle days are numbered, as surely as the horse-drawn carriage days were numbered back at the turn of the previous century. EVs are cheaper to operate, easier to maintain, and more fun to drive.

EVs consist of better tech. Any questions?

Oh. I see that you still have a few.

TL;DR: Your Top Questions about Electric Cars, Answered

If you're looking for one section of one chapter that will make you feel like you know precisely what you're talking about — in other words, if you're looking for a section that leaves you with a truly Dunning-Kruger confidence in your subject matter expertise, similar to the person who glances at a news headline and then heads straight for the Comments section to land a few truth-haymakers — then this, my friend, is the section for you.

TIP

Another fantastic technique for convincing others of your informational bona fides during casual conversation is to mention that you saw a TED Talk. TED Talks serve as a kind of TikTok for rigorous academic study. If you say you watched a TED Talk, first, no one really questions whether you actually did. Second and more importantly, you're now speaking with the borrowed authority/credibility of whatever expert was giving the TED Talk, so whatever comes out of your mouth after "I saw a TED Talk on this. . ." now has the imprimatur of a TED speaker, and you can't exactly invite yourself to speak as one of those things, can you? (You can.)

It's a fantastic technique, but unnecessary for your next EV conversation, during which you will confidently preamble with this statement instead: "So I was reading this book on EVs by Brian Culp, and . . ." — boom!

Alternatively, this is the section you can send your skeptical neighbors and coworkers to whenever they start asking you *the exact same questions that are listed here*. Just roll your eyes and hand them a copy of this book with the bookmark set to this page.

Oh, and besides their utility for impressing friends, neighbors, and coworkers, answers to these common EV questions serve as a preview for the rest of the book.

Here's the list (in section format), in no particular order.

How long does an EV take to charge?

This is *the* single most frequently asked question from people considering an electric vehicle. And people who own an electric vehicle usually provide the same answer, which sounds like the kind of backtalk that might get you a sharp rap on the knuckles if we lived in an earlier era:

> **About 8 seconds**

Kind of what a Brit might call a *cheeky* answer, but also an honest one. Eight seconds is about how long it takes to walk over to the charging cable in your garage after wrapping up your day's activities and plug the car in. The next morning you awaken to a charged battery.

If it sounds like charging your phone on your nightstand, that's exactly what it is. How much you agonize over your phone's charging time? Put your phone on the charger. Wake up to a full battery. What do you mean, "How long does it take to charge a phone?"

Now, I do acknowledge that people asking this question are also thinking about the occasional road trip. As I demonstrate in Chapter 7, road trips represent a *very* small percentage of vehicle use. However, road trips represent a large percentage of people's concern about charging electric vehicles. During road trips, EV owners charge at DC fast-chargers, which are capable of charging most EVs from 10 percent to 80 percent in about 20 minutes. So, 20 to 30 minutes while on a multistop road trip is a general rule of thumb when on the road.

EVs can't get me where I need to go?

Was that a question? Just because a sentence has a question mark doesn't mean it's a question. Fine, let's pretend it was, or at least for the purposes of setting up something for me to argue against.

As I explain in greater detail in Chapter 4, and again in Chapter 13, the average American drives fewer than 50 miles per day. (You can look up the number if you're into that sort of thing: `www.fhwa.dot.gov/policyinformation/statistics/2019`.)

So the question to the non-EV-driving Mr. Strawman is this: Where all do you need to go in a day?

The median range of an EV in 2020 was 259 miles on a full charge, which should be a plenty of range to get you from San Francisco to Lake Tahoe, or to enable the average commute five times before having to recharge the battery.

True, how much range you get can be impacted by conditions like cold weather or terrain — driving from San Francisco to Tahoe is mostly uphill — but the really cool thing about an EV is that, for the most part, you don't drive it around until you empty the battery the way most people empty a tank of gas. Mr. Strawman plugs in the car at night and then wakes up to a full 259 miles of driving range.

Aren't EVs expensive?

Yes. New electric cars are expensive.

Also, new ICE *cars* are expensive.

The average cost of a new vehicle in 2022 in the United States is over $45,000. The real question is, which car — the EV or the ICE — will have the lowest total cost of ownership (TCO) over the time you'll own it. As you'll see in Chapter 4, the EV is likely to be the cheaper transportation option when compared to a similar ICE vehicle.

How long until I have to replace my EV's battery?

In most cases, never. Today's battery packs are engineered to last the functional *lifetime* of an electric vehicle. It's not uncommon for vehicle manufacturers to warranty their battery packs for 8 years or up to 150,000 miles of use.

In a word, most EV owners don't have to think much about the battery pack. Beyond that, we'll discuss best practices for getting the most out of your car's battery in Chapter 8.

Why can't I find somewhere to charge my EV?

This question is patently, almost laughably, absurd. That said, the claim usually comes from a place of misunderstanding, of thinking about charging stations the same way people think of gasoline pumps.

And, if that is your mental model of how all this works, you'd be excused for noticing that there are indeed far fewer public charging stations than there are gas pumps. For now.

This situation is changing rapidly: you can now find over 200,000 public charging stations across the United States, like the ones lining the street where I live. And if that's not the case where you live, just give it another few months. Smart businesses will use charging to lure customers or employees. (It used to be difficult to find a coffee shop that offered free Wi-Fi.)

REMEMBER

Electric charging infrastructure is literally *everywhere*. If you can plug in a lamp, you can plug in a car. (See Figure 1-12.)

FIGURE 1-12: EV charging stations — the free coffee shop Wi-Fi of the 2020s.

What happens if the battery runs out of juice while I'm driving?

What happens when your electric vehicle runs out of battery is the same thing that happens when your ICE car runs out of gas. What's more, just like the range estimators in gas-powered cars that tell you how many miles remain until the fuel tank is empty, electric cars have software that keeps tabs on the battery, telling you how far you can go on the current charge.

That said, there's an important paradigm shift that I think you'll find helpful if you begin your electric journeys thinking of your battery pack in the same way as a tank of fuel. I talk more about that topic in Chapter 7.

Aren't EVs worse for the environment than gasoline cars?

In a word, no.

Whether or not this is a good faith question in the first place, asked by people who are rationalizing their continued use of a product that is polluting their kid's air when they can choose affordable alternatives, is an argument for another day.

The bottom line is that the "EVs are *actually* worse" theorem has been thoroughly debunked. In the first place, EVs are zero-emission vehicles — they have no tailpipe.

"Yeah, but your electricity comes from coal," goes the counterargument. Though this statement may be true, studies have shown that even EVs using coal-generated electricity have a lower CO^2 footprint than ICE vehicles.

What's more, there are many ways to generate electricity, including ways to ensure that your family's entire electric consumption comes only from solar or wind.

"But the batteries!" The argument here is that creating the electric vehicle has a greater environmental impact than the ICE counterpart. This argument would hold some weight if the car's lifetime were only about 18 months. The fact remains, however, that over a car's (either ICE or EV) average lifetime of almost 8 years, the EV emits far fewer greenhouse gases from cradle to grave.

And, as I point out in Chapter 8, the batteries can be recycled.

» **Torquing: Not a new dance move**

» **Braking regeneratively**

» **Miniaturizing the drivetrain**

» **Managing OS upgrades**

Chapter 2
The Fascinating Tech That's (Sort of) Under the Hood

This chapter was written for your inner nerd. You know the one I'm talking about — the one who understands references to *Star Wars* canon, like the one I'll drop later into Chapter 6. It's the nerd who has an extensive Funko Pop collection and framed *X-Men* and *The Incredible Hulk* comic books from the 1960s and who could, if asked, record a 30-minute extemporaneous video essay about the vexing seventh season of *Game of Thrones.*

Your inner nerd is the one who has *this* clock in their office. (See Figure 2-1.)

You know the one: the nerd (fun fact: the actual name is Dummies Man) who's featured on the cover of this and every *For Dummies* book in existence. (Surely it has occurred to you, too, that the triangle-headed character on *For Dummies* covers looks more like the archetypical nerd than a dummy. But what does a dummy look like, anyway?)

But nerding out is awesome! Nerds are the ones who are curious about how the world works, who have the imagination to dream up ways that the world can work even better, and who have the smarts and the perseverance (although it's probably more accurate to say they *develop* those skills) to see those dreams through to reality.

FIGURE 2-1: That's some *gangster* nerd cred right there.

Nerds put together the original electric vehicles in their garages because they had some batteries and wheels and lots of free time on their Saturday afternoons. Nerds do cool stuff with their lives rather than play golf. (Nerds would study aerodynamics and invent better golf balls, I suppose.) Nerds learn about torque and instant power, and about the First Law of Thermodynamics (conservation of energy) that applies when using regenerative braking, and about how stators and rotors work together to turn electromagnetism in to motion.

So tell your inner nerd to start brewing the Dalgona coffee and put on some binaural music, and let's look at the fascinating technology that makes electric cars a better solution for getting from place to place.

Recognizing That Induction Motors Mean Instant Energy

As everyone knows, the first electric motor was invented in 1835 by the famed electricity pioneer Nikola Tesla.

Now, the only problem with the preceding sentence is that it's completely false. Tesla was certainly a giant in the electric motor field, and his contributions were essential to the vehicles we drive today, but I'll get to that in a bit.

Work on converting electrical energy to mechanical energy actually began as far back as the 1740s, in the three-car garage of the Scottish monk and scientist Andrew Gordon. That early work was later built upon by scientists such as Joseph Henry and Michael Faraday.

Even if you haven't spent recent months researching electric motor history (you have enough on your plate with keeping track of the interconnected timelines of the Marvel cinematic universe), you likely recognize the Faraday name from its use by the EV start-up Faraday Future. Michael Faraday's work on the underlying principles of electromagnetic induction throughout the early 1800s was foundational to the invention of the electric motors that would eventually power cars.

TECHNICAL STUFF

Michael Faraday was such an influential figure in the field of electromagnetism that Albert Einstein was said to have kept a picture of Faraday in his study to serve as muse.

The first battery-powered electric motor appeared in 1834 and was the work of a Vermont inventor named Thomas Davenport. Davenport's electric motor was used to power a small-scale printing press, which, the last time I checked, are not cars.

What's all this Tesla stuff, then, if all these other folks were working on electric motors? Surely there's a reason that not one but *two* electric vehicle companies use his name. I'm sure if Nikola Tesla's middle name were well-known, there'd be yet *another* EV start-up using that one on company letterhead. As it turns out, Tesla either had no middle name — Tesla was Serbian and apparently middle names weren't commonplace in Serbia — or if he had one, he never used it.

In fact, there is a reason Tesla and electric cars are now synonymous. In 1887, Tesla invented the AC induction motor, a feat made possible in large part by the financial backing of George Westinghouse — he of Westinghouse Electric fame.

This initial invention was tweaked for the next five years until the first commercially viable induction motor appeared in 1892. What made this motor noteworthy was the spinning rotor, which was the first time an electric motor became suitable for automotive use. (See Figure 2-2.)

Then, in 1896, Thomas Edison's General Electric reached a licensing agreement with Westinghouse (they were either frenemies or rivals, depending on who is telling the history). GE then began development of a 3-phase induction motor utilizing the bar winding rotor design. The new motor design called for 3-phase motors using three alternating currents of the same frequency — a full exploration of which is beyond the scope of this book. Just know that a 3-phase motor transforms electric power into mechanical energy, a phenomenon I cover in a bit more detail in the next section.

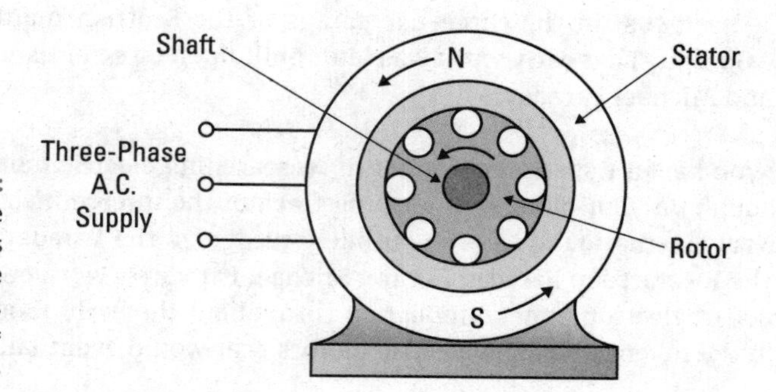

FIGURE 2-2: Stator is the housing. Rotor is the squirrel cage that rotates. And rotation = motation.

Now, more than 100 years after Nikola Tesla's world-shaping invention, electric motors are used to power machines like refrigerators, power tools, lawn mowers, wheelchairs, industrial fans, pumps, and golf carts. (It's yet another instance where the exacting nerd world of inventing things intersects with the boring world of golf.)

And, unlike Thomas Davenport's electric motor, the one from Mr. Tesla is used in electric vehicles.

TECHNICAL STUFF

If you're curious about the rivalry between the Thomas Edison invention factory that was General Electric and the Westinghouse/Tesla partnership, you can nerd out on no fewer than three books: *Empires of Light: Edison, Tesla, Westinghouse, and the Race to Electrify the World*, by Jill Jonnes; *The Electric War: Edison, Tesla, Westinghouse, and the Race to Light the World*, by Mike Winchell; and the novel *The Last Days of Night*, by Graham Moore. I've not read the first two but quite enjoyed *Last Days of Night*, which reads like a mystery/suspense novel. It was named one of the best books of the year in 2017 by both the *Washington Post* and the *Philadelphia Inquirer*.

Seeing What Happens When You Hit "the Gas"

So, what actually happens when you put the ol' pedal to the metal? (We in the electric car business typically don't call it a *gas* pedal, as there's obviously no gas in the car. We usually call it the *accelerator* pedal, or just the accelerator.)

In a single sentence: An electrical field interacts with a magnetic field, which results in a spinning rotor. That's the electric-to-mechanical conversion at work. This spinning rotor then provides the torque that moves the car. (Refer to Figure 2-2 — the one highlighting the rotor.)

To expand on that point just a bit without the need for an electrical engineering degree: Whenever you step on the accelerator, an electrical circuit closes, which allows electrons to move along a wire (specifically, the three conductive wires in a 3-phase motor). Those electrons generate an electromagnetic field. As with all other electromagnetic fields, the field in an electric motor has both a positive and negative polarity.

Now, think back to when you were a young nerd playing with magnets that had fallen out of refrigerator tchotchkes. The poles either attract or repel each other. Essentially, the same thing is going on inside an electric motor, wherein a set of magnets is mounted to the rotor and another set is in the housing that surrounds the rotor. The interplay of the magnetic fields spins the rotor.

Another significant factor in this interplay is the fact that electric induction motors rely on alternating current (240 volts AC), which changes the polarity of the magnetic fields. (Recall that this was Tesla's main contribution to the use of electric motors to applications like cars.) Without alternating current, the electromagnets would simply lock the positive and negative poles together, and no rotational energy would be produced — no spinning of a rotor and no resulting spinning of an axle or a wheel. In short, it's the polarity switching (made possible via alternating current) that keeps an electric car motor spinning.

Most Nikola and Tesla vehicles, along with almost all other EVs, use a variation of Nikola Tesla's 1887 AC induction motor. Today's most commonly used AC induction motors are known as permanent magnet synchronous motors, or PMSMs, although there are subvariations thereof. Like the brushless DC motor, the PMSM features permanent magnets on the rotor and stator. To the untrained eye (i.e. me), a brushless DC motor and a PMSM AC induction motor look almost identical.

With the AC induction motor, electricity flows through windings in the housing, or stator. As this happens, the windings generate a rotating field of magnetism. As these magnetic lines pass through perpendicular windings on a rotor, they *induce* an electric current. That's the induction part of an induction motor.

In any event, it's this interplay of rotating magnetic forces that causes the rotor to turn. One could even say that the rotor was *induced* to turn, and were one to say that, one would be correct.

YOU DON'T *NEED* A TESLA

Or at least you don't need the work of Nikola Tesla and the AC induction motor in order to run an electric car.

The very first electric motors that powered cars, back in the early 1900s, like the one that killed Henry Bliss (for more on the untimely end of Henry Bliss, see Chapter 3), were actually DC motors that used brushes. These "brushes" — conductive flaps of metal in a stator — essentially provided the same function as AC power in AC induction motors: It turned electromagnetic fields into rotation. (If you've ever seen sparks through the motor cooling vents on something like a power drill, you've seen a DC motor brush at work.)

There are also "brushless" DC motors, similar to AC induction motors in that they place permanent electromagnets on both the rotor and the stator. Then, by using an external controller to alternately switch the field windings from positive to negative, the brushless DC motor creates the rotating magnetic field. In fact, brushless DC motors are commonplace today and are even used in some very light-duty locomotive purposes. Both my electric scooter and my electric bike use a brushless DC motor.

Getting further into the weeds about brushed-versus-brushless electric motors is, alas, beyond the scope of this book. My inner nerd sends its apologies.

TIP

Further reading: Wait — what? Why read when you can watch videos instead? Seriously, the animations in videos these days do what words and static images simply cannot. So, for further care and feeding of your electric motor nerd, I send you to these resources on YouTube:

- **How Electric Motors Work — 3 Phase AC Induction Motors:** `www.youtube.com/watch?v=59HBoIXzX_c`
- **How Does an Induction Motor Work?** `www.youtube.com/watch?v=AQqyGNOP_3o`

As mentioned, when you can get a rotor to turn, you're on your way to turning electricity into torque. And torque is a very good thing.

Comparing Torque and Horsepower

I'll end up explaining this whole torque versus horsepower thing in more detail in Chapter 10, but let's do a quick tour of the concept here.

And quick is the operative word. As in, electric cars are quick. Staggeringly quick. As in head-pinned-against-the-headrest quick. As in maybe-think-of- packing-a change-of-underwear quick.

As in quicker than anything you've ever experienced on four wheels.

Why? Because of torque. And also horsepower. But the quickness — that visceral feeling of acceleration you get when launching from a standstill — comes from the *instant* torque delivered thanks to an electric induction motor.

Torque measures the rotational force your motor (or ICE engine) delivers to the car's axle. (For this reason, torque is also very significant when towing.)

Horsepower, on the other hand, measures how much power is being transferred from the motor/engine to the wheels.

They sound very similar, and they both contribute to driving performance. What you feel when stepping on the accelerator is a combination of both forces, although each works in different ways to get the car going. In essence, torque is the force that gets your car to initially accelerate. Horsepower's job is to get up to speed and maintain that speed while driving.

Again, both are needed if you want a car to go fast (well, they're both need to get a car to go *anywhere*), but torque is the stuff that gets your car moving from a dead standstill (it also helps instantly launch the car from say, 20 to 60 in the blink of an eye). It produces the "EV smile" that so many experience when feeling that initial burst of speed.

And experience it you must.

On my first test drive of an electric vehicle, the consultant advised me not to exceed 100 mph. I laughed at that suggestion — at the notion that I would in fact hurtle down a highway at 100 mph during a test drive.

The advisor did not.

Put simply, electric cars are some of the fastest cars on planet earth. In terms of straight line 0–60 or 0–100 times, they are the fastest production cars, full stop. (Or at least production cars that don't cost over $1 million. Does anyone really consider a $1 mil car a production car? No.)

In fact, here are three of the very, very quickest and fastest quarter-mile production cars in existence. See if you can spot a trend.

Tesla Model S Plaid: 1020 horsepower

0-60: 2.1 seconds

Lucid Air Dream Edition Performance: 1111 horsepower

0-60: 2.5 seconds

Porsche Taycan Turbo S: 616 horsepower

0-60 2.6 seconds

Yep: insanely fast, and yet not a tailpipe in sight. Again, please stop by Chapter 10 for a more detailed breakdown of the fascinating physics that makes all the speeding up possible.

Speaking of which. . .

Letting Physics Do the Braking

It takes a lot of energy to move a vehicle that weighs 2,000 to 3,000 pounds down a road at 55 miles per hour. Specifically, humans have to figure out how to harness either the power of a small explosion (the fuel for which is more powerful than dynamite — see Chapter 3 for more info) or enough electrical power to run an entire house for several days.

Knowing how EVs make energy while driving

The thing about generating the energy needed to overcome this much inertia — that whole deal about "A body at rest stays at rest, especially if that body is a 2,000–3,000 pound car" — is that all that energy is wasted every time you use the car's brakes.

In case you're a physicist (do physicists pick up books with *For Dummies* in the title?), you're probably already reaching for the keyboard to correct me: Fine. The energy isn't technically *wasted* — it's instead *converted* from kinetic energy, which is the energy of something in motion, into thermal energy, which I guess I should explain (in case you're a golfer) is heat. Unfortunately, heat won't do anything to get your car moving again, so it just dissipates into the atmosphere.

HEY, WHAT'S YOUR DEAL WITH GOLF?

Nothing in particular, other than it's a boring way to spend an afternoon, it wastes obscene amounts of space that in most cases might otherwise be turned into lively, walkable neighborhoods in a country that's desperately short of affordable housing, it wastes obscene amounts of water, it requires entirely too much fossil-fuel-derived fertilizers just to have a place to "play," and it requires an obscene amount of money to even have a *chance* to develop the skills necessary to compete at the professional level. As with most of the US, a meritocracy it is not, although this is not the tale spun by its supporters. In this way, golf reminds me too much of polo, water polo, and yacht racing.

I do enjoy the time with friends during a round, and I'm actually a decent golfer, given how little I play — it's a ball-and-stick thing that carries over from a lifetime of playing baseball — but I get either twitchy or sleepy by about the 11th hole. Finishing an 18-hole round is torture for me. All told, I view golf as a pursuit that needs to go the way of bull-fighting or English foxhunts — neither has much place in modern society.

And golf *requires* internal combustion engines in its current iteration, where lawn mowers and the transportation of material to-and-from are necessities. Until that changes, and until golf rounds for amateurs are 10 holes (or maybe 12), I would rather spend my weekends on my bike, hiking in a park or forest, or writing sentences that have never before existed rather than wandering around a golf course in search of a wayward tee shot.

So, yeah — besides the foregoing, my deal is nothing in particular.

Wouldn't it be better, then, if there were a way to capture and store all that kinetic energy so that it can be used to move the car?

It's almost as though that question serves a segue to. . .

Seeing what happens when you let up on the "gas"

Electric vehicles use a kind of braking system that recaptures much of the kinetic energy generated by the vehicle when it's in motion. This kinetic energy is then converted into electricity, which recharges the battery pack.

The name of this braking system used by cars with electric drivetrains — and this includes BEVs, PHEVs, and HEVs — is *regenerative braking* (or *regen*, for short).

Although regen in EVs has been around since the first Toyota Prius rolled off the assembly line in 2002, the concept did not coincide with the advent of electric vehicles. In fact, the first use of regen was on trolley cars, like the ones that still traverse the hills of San Francisco. It takes a lot of energy to go up a hill; and you can recapture some of that energy when you're descending.

So then: Remember how I just discussed that the key to an electric motor is a spinning rotor? What's interesting about this concept is that spinning a rotor is the same process that's used to generate electricity at scale. In the instance of a power plant based on coal or natural gas, the burning of fossil fuels creates heat, which creates steam, which spins a turbine, which in turn generates electricity.

Regenerative braking does this at a smaller scale, turning the electric motor (and its spinning rotor) into a miniature power station.

Pretty cool, huh?

And also, pretty efficient.

What it means for the performance of the electric vehicle is that the car's kinetic energy is used to extend the life of the battery — it increases an EVs range.

That said, regen breaking isn't a perfect closed-loop system. The best regen braking systems still lose around 10 to 20 percent of the energy being captured, and then the car itself loses another 10 to 20 percent efficiency when converting electricity to motion, as described here: `www.tesla.com/blog/magic-tesla-roadster-regenerative-braking`.

Add it all together and what you get is a system that recaptures roughly 70 percent of the kinetic energy lost during the act of braking — energy that can then be used again when accelerating the car.

Terrain also plays a significant factor in regen. If you're headed mostly downhill for long stretches — say you're coming back from Lake Tahoe to Sacramento along Interstate 80 after a weekend in the mountains — you can expect to barely dip into the battery's range, because the car undergoes several long "braking" periods (even though the brakes are never applied) in order to maintain speed while travelling downhill.

If you're headed the other way — from Sacramento to Tahoe — you don't regenerate much while going up a hill. You instead just slow your literal roll, thanks to gravity.

REMEMBER

When weighing an electric car versus an ICE, keep in mind that the EV is way, wa-a-ay more efficient with its energy. ICE cars can convert only about 20 percent of the gasoline burned into usable energy that propels the car forward. (Diesel engines tend to be a bit higher.) As you might expect from a machine that relies on tiny explosions, most of the energy produced in an internal combustion engine turns into heat. And none of *that* energy can then be recaptured. It, too, is lost as heat as brake pads rub against brake rotors. Thought of another way, driving an ICE is an excellent way to turn energy into (brake) dust.

And, for the driver of the EV, regen means that if you drive with reasonable caution, you can complete most of your trips with minimal braking, or at least very little of the braking that requires the brake pads. In sum, BEV drivers enjoy virtually brake-free driving, and they see an increase in range as a bonus.

Under your foot, this translates to the one-pedal driving experience. When driving an EV and anticipating a stop, all you have to do is take your foot off the accelerator. The EV then decelerates, which is the regenerative braking in action. Unless the accelerator is pressed again, regen brings the car to a full stop.

TIP

If you want the EV braking to behave like a traditional ICE vehicle, you have ways to disable regen so that you just coast every time you let up on the accelerator. Don't do this, though. Don't just wear out brake pads for no reason. What are we — Neanderthals?

In a cool bit of feedback, most vehicles give the driver a visual indicator that energy is being recaptured. In the Lucid, it looks like a speedometer, as shown in Figure 2-3.

FIGURE 2-3: A Lucid regen indicator.

In the Tesla, it's a line across the top of the display.

In fact, most of the time when stopping an electric car, you have to apply the "gas" when approaching a stoplight or stop sign as you finesse the regen. It takes a little getting used to, but it's nothing that most drivers don't conquer over the course of two or three drives. I can tell you from my own experiences and the reports from others who have switches that any time you get behind the wheel of an ICE again, the braking just feels . . . wrong.

TIP

This advice should be rather obvious, I guess, but I'll add it anyway: When stopping for some kind of emergency — a moose running out in the road or something like that — don't wait for regen. Slam on the brakes.

Making More Room for the Fun Stuff

Even though electric induction motors are marvels of engineering, relatively little attention is paid to them, even by the auto manufacturers themselves, especially when compared to way they're positioned in the marketing of ICE cars. At the very least, almost every advertisement or spec page of an ICE vehicle mentions whether the vehicle runs a 4-cylineer or 6-cylinder engine. More powerful cars and trucks feature their engine characteristics front and center: HEMI, DOHC, V8, twin turbo, and liter displacements are terms that I've heard while watching halftime ads, even though I have no idea what they refer to. I guess a 6.2-liter engine is better than a 4.4-liter engine. Or does it depend?

In any event, astute car shoppers may notice that the Tesla Motors Model 3 web page features a specs section but makes no mention of the motor. Likewise, the Chevrolet web page about the Bolt EV only mentions an electric drive unit. The word *motor* does not appear.

Maybe it's a bit of an inferiority complex with electric motors, because they simply take up far less space than the motor needed for a gasoline-powered car. I don't know. I can't read an electric motor's mind.

Engine is the term for a device that converts fuel to mechanical energy. *Motor* is the term for a device that converts electricity to mechanical energy.

What I do know from looking up terms online is that internal combustion engines take up a lot more volume than an electric motor. (Is that what the liter thing refers to?) I can give you several pieces of data to back up this statement, but all you need are your own two eyes to confirm. Go to just about any auto club meet-and-greet and you'll be welcomed by men (and, yes, they are mostly men) eager to show off the massive size of their engines.

Of course, internal combustion engine volumes vary widely, depending on whether the engine is in a Ram 350 pickup truck or in a Honda Fit, but you get the idea. Everything under the hood is devoted to storing the engine block, 12-volt battery, and tank of antifreeze.

Meanwhile, pop the hood on almost any electric vehicle and you're treated to the sight of extra storage space, as seen in Figure 2-4.

This beautiful and extraordinarily practical storage space exists because electric motors are many times smaller than internal combustion engines. In fact, the motors aren't even mounted under the hood, like IC engines are.

Instead, most EV motors are mounted right where they deliver the power — adjacent to the vehicle axle — almost always along the rear axle, in fact, and in many cases next to both the front and rear axles, as is the case with the Lucid Air, Plaid Model S, Porsche Taycan, and several others. If the EV is an all-wheel-drive car, it will have at least one motor per axle. (See Figure 2-5.)

The real question, though, is what to do with all that extra space.

FIGURE 2-4: You know what (just barely) fits in here My giant duffel bag of CrossFit gear.

Lucid Motors

FIGURE 2-5: Remarkable. These two miniaturized motors make this car faster than a Ferrari.

CHARGED

Knowing What to Do When There's (Almost) Nothing Under the Hood

Electric-vehicle engineers have begun finding all sorts of cool uses for the extra space afforded by not needing an engine block.

At Lucid Motors, the engineers I work alongside (read: those who tell me things using simplified language) have even built a design framework around this smaller drivetrain. It's called the Space concept. According to them — well, to us — the *Space concept* maximizes interior cabin space, offering drivers and passengers the interior square footage of a Mercedes S Class while having the exterior wheelbase of a Mercedes C Class. (See Figure 2-6.)

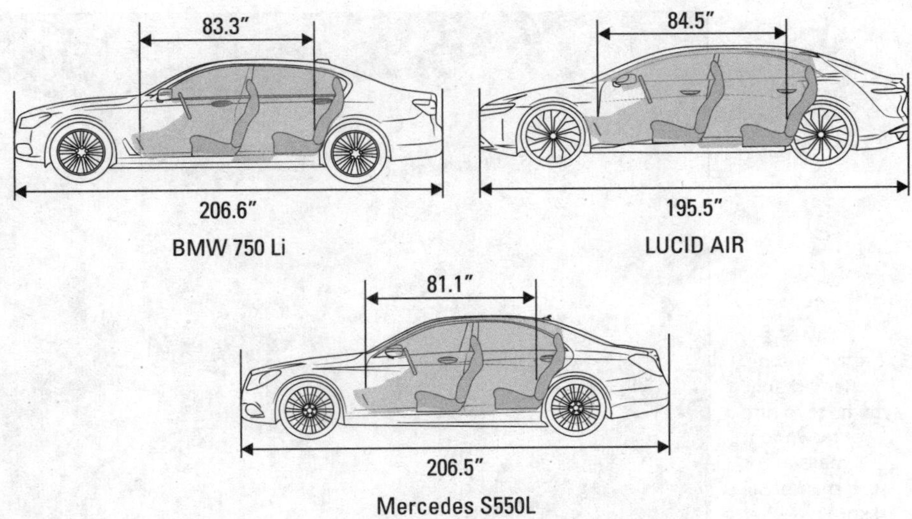

FIGURE 2-6: An electric drivetrain means a compact sedan with a spacious interior.

The Air gives back much of the space savings from a pair of small electric motors, in the form of the largest frunk available in an EV. *(Frunk = front + trunk.)* And some models let you do what you see in Figure 2-7.

But of course Lucid isn't the only company finding other uses for room freed up by the electric drivetrain. Rivian, for example, has bored a storage "tunnel" between the passenger cabin and the truck bed. (See Figure 2-8.) Such a tunnel is perfect for tucking away items like luggage for your camping trip, tools for your worksite, or even an entire camp kitchen! Like, a two-burner grill and everything!

Meanwhile, our friends at Tesla, GM, Ford, Hyundai, and Toyota all have thought up clever uses for the space normally given to mufflers, gas tanks, and the like.

For example, the Model 3 Tesla and the Model Y Tesla feature an extra cargo area that's hidden under a cover. (See Figure 2-9.) When we drove our Model 3 across the country, this additional storage is what we used to transport some of our most precious cargo during our move from Kansas City (our stock of good whiskey, along with some barbecue sauce from Joe's KC and Jack's Stack). It's such a significant space, in fact, that several manufacturers offer organizers just for that space, as seen in Figure 2-9.

The Ford Mach-3 has a similar underfloor space, which gives it almost the same amount of cubic feet of storage as the much taller Escape SUV.

FIGURE 2-7: Executive seating. Remarkably, you have to hire your own foot masseur or masseuse. I know — what a rip off!

Lucid Motors

FIGURE 2-8: Remarkably, you have to hire your own chef to cook a meal. What a rip off!

Roschetzky Photography / Shutterstock.com

FIGURE 2-9: A case of Weller and three bottles of sauce is the answer to your question.

EVANNEX

The Hyundai Ioniq 5 includes a generous frunk space, like most of its electric vehicle kin. And the Chevy Bolt should prove more than up to the task for the storage of hockey gear, a dog or two, and whatever else people buy in bulk at Costco that they probably shouldn't.

Driving an iPad with Wheels

You know how this works: You plug in your iPhone or Android device at night and you wake up to a brand-new phone, or at least very often receive several brand-new features.

So too with (most) electric cars.

Almost every ICE vehicle being released now is as much a rolling piece of interconnected software as a piece of machinery. Software now very often controls features such as throttle response and suspension stiffness, not to mention the myriad of features controllable by way of the car's infotainment system.

The electric vehicle, then, takes this trend and turns it up to levels that would make Nigel Tufnel proud. (Possible that you're a *This Is Spinal Tap* fan, yes? If not, Google is your abiding friend.)

Faster and further — all by changing a few words

Throughout their history, cars have not afforded us the capability to improve themselves the way that our phones and tablets do. Typically, the best experience an owner will have with an ICE car is when the car is brand-new — the vehicle is a machine of steel and glass, and you get what you get. Yes, you can replace or add things such as radios (hello, aftermarket Alpine sound system in my '82 Civic), wheels, and even the engine, but that's not technically the same thing you drove off the lot. (In fact, such a modification of original parts is suggestive of the famous thought experiment about the Ship of Theseus.)

Over time, the bells and whistles that seemed so modern at the time of purchase become stale, if not obsolete, as manufacturers continue their march of progress. That awesome 8-track player (hello, '78 conversion van with the custom paint job) gives way to the cassette player, which makes room for the CD player, which gives way to the USB port, all of which are rendered virtually useless by the Bluetooth connection and unlimited data plan.

All this changes with over the air updates. No, it can't change the physical car components. But yes, it can improve the actual on-road performance of those physical components, as how those components behave is often a function of software.

For instance, imagine that simply by letting your 2018 BMW 5 Series install updates automatically, the way your MacBook OS or your copy of Microsoft Office does, you wake up the next day to a Beamer that gets better gas mileage. Or added horsepower. Or improved, advanced driver-assistance system (ADAS) features when driving on the highway.

I don't have to imagine any of these scenarios because our 2018 Model 3, which would certainly be considered a competitor to the BMW 5 (and 3) Series of sedans, received *all* these improvements while the car was in the garage and connected to Wi-Fi.

Over-the-air updates have even been used to address National Highway Traffic Safety Administration (NHTSA) recalls.

In fact, going back to our comparison of Tesla and BMW, a recent analysis of public data showed that BMW issued 60 recalls between January 2020 and January 2022, all of which were addressed with physical fixes. Meanwhile, 7 of the 19 Tesla recalls issued during the same period (nearly 40 percent) were resolved with a software update.

Using an over-the-air update to address a safety issue (or at least what NHTSA deems a safety issue) confers several advantages. One is that they're far less costly in terms of time and money than a trip to the repair shop. Another significant upside is that remote updates can all but guarantee that all affected vehicles receive the fix.

REMEMBER

Though most safety recalls earn the safety moniker, some don't. For example, while I wrote this section, the NHTSA recalled 54,000 Teslas using the Full Self Driving (FSD) beta features because the cars didn't come to a complete stop at stop signs. They rolled through stops at 2 mph if the coast was clear, putting these FSD cars ahead of 99 percent of all human drivers in terms of behavior at a stop sign with no cross traffic. Nevertheless, the NHTSA issued a recall, and Tesla rewrote the FSD code so that the cars would come to a complete stop. Yes, this did affect my family's Model Y. (I'm the source of this info, but here's an alternative one: `www.reuters.com/business/autos-transportation/tesla-recalls-nearly-54000-us-vehicles-rolling-stop-software-feature-2022-02-01`.)

Over-the-air updates deployed into today's electric vehicles are divided into two categories:

- Firmware updates
- Software update

Either a few weeks from now, or a few months from now, when you look down at your phone and see that your EV is ready for an update (you'll receive a notice in the car's app most likely), you should probably be more excited about the firmware update. Why? Because firmware updates affect the software governing vehicle behavior. This can mean increases in range, acceleration, and other characteristics that affect driving. A firmware update is analogous to getting a new operating system for your phone, like going from iOS 14 to 15.

The software update, on the other hand, typically happens more frequently than a firmware update, and includes features like improvements to existing features or bug fixes meant to address issues left over from previous software updates.

Again, the best part of all of this is that these improvements to your vehicle happen automagically, just like they do for your phone. You receive a notification from your car's connected app; that notification usually includes an option to install now or schedule the update, and the update then occurs at a time most convenient to you. Nine times out of 10, it happens while you're fast asleep.

TIP

You know how you can't use your phone while it's in the middle of an over-the-air update? Same goes for the car. While a software upgrade is taking place, you're notified that the car will be undrivable during this operation. For this reason, it's not recommended that you initiate a software update shortly before any planned travel. Most cars do have some way of interrupting an in-progress update to make the car drivable in case of an emergency.

What exactly might this look like?

Who knows? The fun part is not knowing what goodies might be coming your way. It's waking up one morning and being able to play *The Witcher* on your car's infotainment system.

To give you some kind of indication of what might be in store, however, here are just a couple of noteworthy examples of what over-the-air software has delivered:

- In 2019, Tesla pushed Dog Mode which runs either the heater or air conditioner in order to keep the car at a comfortable temperature while the car owner is running a quick errand. The Tesla display then displays a message letting others know that your pet is perfectly safe.

  Tesla has been providing OTA updates since its model S sedan in 2012, so Dog Mode is just one of many features that have been provided for free to Tesla owners. Other examples include Camp Mode, Sentry Mode, blind spot visualization improvements, and the previously mentioned performance and safety updates. The full list would take several more pages to detail.
- In 2021, Volvo (Polestar) released an update that increased the Polestar 2's driving range and charging speed and deployed an update to the Android Auto operating system that controls the infotainment system.
- In 2022, Lucid updated its infotainment software to add support for the Alexa voice assistant, Apple CarPlay, and navigation enhancements like *waypoints,* which are stops along the route.
- In 2022, Hyundai pushed its first OTA update for the new IONIQ 5. The update improved the infotainment and map software.

Predicting the future of OTA updates, and the car as a computing platform

Many of the improvements to car software and firmware have been released as free updates — a perk of being on the forefront of technology. But for an idea of where this is headed by about mid-decade, I ask you to consider what product on the planet has the highest profit margin.

Pharmaceutical drugs?

Maybe, but I was thinking of software.

Most of the software goodness coming to a car near you could end up being a pay-for-play proposition. Once more, we're back at the paradigm of an iPad or Android device. The hardware is the platform to sell (or rent) the software. Here are a few examples:

- **Example 1:** The Mercedes EQS lets you turn the front-facing camera into a dashcam — for a fee. You can put the car into Beginner Driving Mode, which limits the acceleration and power — also for a fee.
- **Example 2:** BMW is now offering ADAS cruise control, adaptive high beams, and other infotainment features from the BMW store, which, like Apple's App Store, doesn't require an in-person visit in order to make a purchase.
- **Example 3:** Tesla sells Premium data connectivity for USD $9.99 per month. (Premium connectivity is necessary for features that require data — stuff like streaming music, live traffic, and more. It's pretty much essential, so just take my darn money.) In addition, you can now access Full Self Driving as a subscription service rather than an up-front purchase of $12,000. As FSD is still not *technically* FSD, I highly recommend the subscription route, using FSD for longer trips when needed. (On the other hand, FSD is part of the car once purchased, so getting FSD is going to result in a higher resale value.)

» Exploring the possibility of a vehicle fire (ICE vs. EV)

» Owning a car that improves over time

Chapter 3
Electric Vehicles and a Safer Drive

What's the most dangerous thing you do every day?

Drive your vehicle.

The amount of death and destruction that occurs on the world's highways and byways is nothing short of a public health crisis. In fact, the numbers tell us that auto crashes are the leading cause of death for people between the ages of 1 and 54 in the United States. (Source: `https://www.cdc.gov/injury/features/global-road-safety/index.html`.)

Looking beyond my home soil doesn't make for a better view. The World Health Organization (WHO) tracks auto accidents as being responsible for 1.3 million deaths every year, more than the global total for murders and suicides combined. And, if you're thinking that this figure is also closely tied to matters of race and class, you're not wrong. Globally, more than 50 percent of fatalities are pedestrians, bikers, and motorcyclists rather than car drivers and passengers, who tend to be wealthier. About 93 percent of road deaths occur in low- and middle-income countries.

And, somehow, the COVID-19 pandemic has only exacerbated this problem — even though statistics show that people were driving fewer miles over the two year period of 2020–21, fatalities caused by automobile accidents (including both pedestrians and vehicle occupants) *rose* over that span. As though dealing with a lethal pandemic weren't enough, 2020 saw the biggest single-year spike in automobile deaths in nearly a century. (You can check out the info at `www.nsc.org/newsroom/motor-vehicle-deaths-2020-estimated-to-be-highest`.) I share a few more statistics later in this chapter, but know in advance that they don't make for pleasant reading.

And yes, I'll allow that there's plenty of room for nuance with my use of *dangerous*. Here, the context should be read as an *imminent* threat to life and limb. Sure, heart disease is the *overall* leading cause of death, so just living is a risky proposition after a certain age, it seems. I can also make a convincing argument that scrolling Facebook or Instagram is more dangerous to your *mental* health, for example, or that smoking is more of a threat to your overall quality of life. Oh, and in the US, people always have guns! The "good guys with guns" manage to mow down *almost* as many people as cars do, despite the fact that few people "own guns" (oops, sorry — the autocorrect feature revised *hoard weapons caches* to *own guns*) while almost everyone has to get to work, buy food, and cart kids to piano rehearsals. So, in terms of the event that has the greatest chance of ending in tragedy *today*, getting in the car and making that grocery run wins? loses? — hands down.

So a vehicle — any vehicle — that reduces the odds of a grocery run ending in catastrophe is certainly worthy of further discussion, and that's exactly the conversation I focus on in this chapter. (Although I'll certainly bring it up elsewhere. Safety is kind of part of the DNA of this entire book.)

I start things off with a quick look at the current reported state of electric-vehicle safety, address specifically the risk of an EV (and an ICE car for that matter) catching fire, and then wrap up with some notes about how EVs are almost certain to improve their stellar safety track record in the years to come.

Why You Might Be Under the Impression That EVs Are Unsafe

Let's begin with a tour of recent(ish) headlines:

- **GM recalls another 70,000 Chevrolet Bolt electric cars**

  `www.cnn.com/2021/08/20/business/gm-bolt-ev-recall/index.html`

» **Cargo ship carrying Porsches, VWs, Bentleys headed for U.S. catches fire in Atlantic**

`www.caranddriver.com/news/a39123584/cargo-ship-porsche-vw-fire-atlantic`

» **Tesla sued over Model 3 suspension failure that allegedly caused a deadly crash**

`www.theverge.com/2022/2/17/22938877/tesla-model-3-lawsuit-suspension-ev-fire-crash-florida`

EVs are new. They're sexy. For now, they're the man-bites-dog angle that draws eyeballs and clicks to a headline. And when it comes to Tesla, you have a man-bites-dog angle combined with a man-who-wants-to-fly-to-Mars-and-also-hosts-SNL-and-agitates-like-a-petulant-teen-from-his-Twitter-account angle that's positively catnip to website and print media publishers — and their readers — across the globe.

(I've already mentioned this point elsewhere, but if you're looking for an author who rails against the *lame*stream media, please either find another book or leave YouTube on Autoplay. People make mistakes, but I unequivocally *trust* the main-stream media.)

For a host of reasons, electric vehicles are drawing well-deserved interest. (Who knows — maybe someone will be asked to write an entire book on the topic one day soon?)

Because they're commonplace, and because they so often play a role in the stories of everyday human tragedy, traditional ICE cars aren't receiving quite the same amount of coverage, especially when you factor in the proportion of ICE-related accidents and fatalities to that of EVs. In other words, no newsroom will spend much time covering a traffic fatality involving a 2017 Buick Enclave, a 2018 Ford F-150, or a 2019 Honda Civic.

And the thing about some of these headlines — this holds true of pretty much all headlines that aren't PhD-dissertation in length — is that context matters. And context is what the article is for. If you read the article, the headline has done its job. If you only read the headline, you haven't done *your* job.

Depending on your favored news sources, if you were to only glance at headlines [see also: Facebook feeds, Reddit, Twitter, and all other online [mis]information sources], you might reasonably conclude that EVs are death traps, right? Sure, EVs will save you time and money, and some can even make fart noises — in exchange for killing you in a grisly car inferno?

Right?

As you will see, not exactly.

As you will (definitely) see, the story of electric vehicle safety is one that tends to paint EVs in a *very* favorable light indeed.

In fact, learning about electric vehicle safety should prove an effective countermeasure to the sense of dread and futility you might feel about what I mention in the chapter intro: Driving a car is *extremely* dangerous.

So let's roll up our sleeves and look beyond the headlines. As a bonus, it's much more interesting when you do.

A Brief — and Instructive — History of ICE Vehicle Safety

As dangerous as traveling in (or around) motor vehicles is today, it has *never* been all that safe. And as bad as the numbers are now, driving in the 1920s was a *lot* more lethal from a percentage standpoint than driving in the 2020s. To wit:

- The first recorded automotive death occurred in 1896, when Bridget Driscoll stepped off a London curb and was struck by a car driven by Arthur Edsall. Despite the vehicle's top speed of . . . 4 miles per hour(!) . . . neither Driscoll nor Edsall was able to avoid the collision.

  In what would perhaps foreshadow the next 120 years (and counting) of a campaign by both the automotive industry and the legal profession to blame pedestrians for being hit by cars, the fatality was ruled an accident and Edsall wasn't prosecuted. (You might know this campaign by its more modern branding — namely, jaywalking or simply "distracted pedestrians.")
- The first pedestrian death in the US occurred on September 13, 1899, when a man named Henry Bliss either stepped out of a New York City streetcar, or was helping a woman do the same, and was struck by a taxicab. He died of his injuries the following day.

  And get this: That NYC taxi was electric — in 1899, no less!

  As with the incident involving Bridget Driscoll, the NYC taxi driver was arrested and charged, but ultimately acquitted. Starting at the turn of the century, killing people who dare to exist, on a street no less, became something that generally just wasn't punished.

WHAT IS A JAYWALKER ANYWAY?

The etymology of "jaywalking" — which means to cross the street without regard to either posted or implied civic law — is interesting. The term "Jay" was used around the turn of the 20th century to insult someone for being ignorant or oblivious.

Specifically, "jaywalker" is a variation of another word used to describe a very specific type of bad driver: a "jay-driver" took their horse-drawn carriage down the wrong side of the road.

The prefix was then popularized by the automobile industry during the 1940s in news-reel films — you've heard of jaywalking but likely not jaydriving — in an attempt to essentially scold people for getting in the way of cars. No one wants to be a jay-*anything*. But the term originally had nothing to do with crossing the street outside of some painted lines (gasp!). It had to do with a lack of sidewalk etiquette during a period where (effectively) no one had to consider being hit by a car.

You might suspect that the term was coined in New York, where pedestrians and carriages regularly competed for space. However, the first known example of the word "jaywalker" in written form was in Kansas, home to the Kansas Jayhawks, in-state rival to my Kansas State Wildcats.

I am an unabashed jaywalker, by the way. I won't tempt fate by stepping in front of a moving pickup truck or anything, but streets are for people, and pedestrians have the right of way. (And I was once "pulled over" for jaywalking, which, yes, I'm including here as a flex. I'm nothing if not self-aware.)

A Short History of Everything (Car-Safety-Related) in 2 Minutes

Well, I guess whether it's limited to 2 minutes depends on how fast you read, but here are the automotive safety highlights spanning the time between Ms. Driscoll and Mr. Bliss to today:

- **1908 — safety equipment (is that a thing?)** The Ford Model T began to be equipped with safety glass (glass that doesn't splinter and break like a bedroom window might, sending long, lethal shards of glass into the passenger cabin).

- **1914 and 1918 — stop signs and stoplights:** The first stop sign appears in Detroit in 1914, and the first three-color stoplight is installed four years later, also in Detroit.
- **1950 and 1959 — seat belts:** The year 1950 saw the first car, the Nash Rambler, with lap seat belts. In 1959, Volvo made itself synonymous with car safety by making shoulder seat belts a standard feature.
- **1955 — driver's education:** Michigan becomes the first US state to require driver's ed to legally operate a vehicle.
- **1966 — vehicle safety regulation:** The US Congress passes the National Traffic and Motor Vehicle Safety Act. Shortly thereafter, seat belts became mandatory equipment, a development that was met with fierce resistance from auto manufacturers.
- **1974 — first airbags:** General Motors manufactures the first car airbags, a precursor to the more advanced collision protection devices that are now standard equipment on vehicles.
- **1985 — standard antilock brake systems (ABS):** Mercedes tries to take the public awareness safety mantle from Volvo by making antilock brakes (on some models) and airbags standard equipment (on all US models).
- **2000 — active safety features:** The decade of the aughts saw the first active safety systems put into production, including automatic braking, adaptive cruise control, and lane-keeping systems.
- **2010 — advanced driver assist system (ADAS) features:** The 2010s introduced many new car drivers to pedestrian detection, parking sensors, backup cameras, blind spot warnings, and night vision cameras.
- **2015 — self driving:** The year 2015 marked the first year that Google's self-driving cars began testing on San Francisco Bay Area streets. (Chapter 11 looks at self-driving in greater detail.)

IF YOU WANT SAFER CARS, ENGINEER (OR RETROFIT) SAFER ROADS

An interesting school of thought, worthy of an entire book or two, maintains that the way to make cars safer becomes mostly a function of making *roads* safer.

More specifically, it's a matter of engineering the roads to prioritize safety over speed. The implication here is that for the past 70 years of road design, street engineers have had it backward: Because of design codes that date back to the 1950s, engineers have prioritized speed over safety when it comes to road design.

Chuck Marohn's *Confessions of a Recovering Engineer* and his *Strong Towns* (both published by Wiley) are two books I devoured as though they were Jack Reacher novels, and, frankly, Marohn's books are more fascinating. Reacher books are fun; Mahorn's changed the way I see the world.

For example, I can't help but notice now that most of the roads in both my childhood hometown of Overland Park, Kansas, and my current hometown of Fremont, California, have more in common with a drag strip and/or an airport runway than streets where human beings should be out walking, shopping, biking, or pushing a stroller. (See the figure below.) It's not comfortable to even be on the *sidewalk* of a road where the posted speed limit may be 35, yet trucks pass by at 50 miles an hour (or more) because road engineers have built the road like a six lane interstate highway, signaling to all drivers that the *appropriate* safe speed is 55 (or more).

In any event, I can't heap any more praise on a book than I have for *Confessions*. You should read it. You should internalize its observations and insights. If you have to choose between this book and *Confessions* because of time and resource limitations, read that one. After reading it, you should try to change the neighborhood you live in one small act at time — doing so will result in a safer, more human place to live and work.

As you can see with this quick tour, *all* vehicle safety has been progressing on a continuum from practically the time of the first recorded accident. And omitted from the discussion here in the interest of time are safety factors such as child car seats and awareness and enforcement surrounding the topic of drunk driving.

The bottom line is that if you buy practically *any* new car today, it's safer than practically any car manufactured 15 years ago, which in turn is much, much safer than anything manufactured 50 years ago. And, as you have the chance to explore next, electric vehicles build on that legacy of steady, incremental improvement and then add some safety advances that are uniquely their own.

Examining Safety Features (and Statistics) for Electric Vehicles

As with much of the talk about modern electric vehicles, the discussion of EV safety begins with the company most responsible for bringing EVs to the attention of a modern car consumer: Tesla.

Our discussion begins in 2013, shortly after the launch of its flagship sedan, the Model S. What happened from a safety standpoint in 2013 that put EVs at the center of that discussion?

It was this: The Model S achieved the highest ratings across all five safety categories measured by the National Highway Traffic Safety Administration (NHTSA).

Attaining a sweep of the NHTSA safety ratings is a very big deal, so it's worth taking a moment to unpack what's involved here.

What's NHTSA's line, anyway?

The National Highway Traffic Safety Administration's stated mission is to "keep people safe on America's roadways." It carries out that mission by a variety of initiatives, like promoting safe behaviors on US roads and highways, collecting and publishing safety-related information such as deaths caused by motor vehicles (such as the numbers quoted earlier), and helping educate the public about risky driving behavior such as distracted driving.

Much of this work is done *out* of the view of the general public. However, NHTSA also includes enforcement within its scope of activities, and these two programs are part of NHTSA enforcement efforts that are *quite* visible to most of the driving nation:

- **Vehicle safety recalls:** Most of the driving public has encountered a recall. NHTSA issues recalls when "either the manufacturer or NHTSA determines that a vehicle, equipment, car seat, or tire creates an unreasonable safety risk or fails to meet minimum safety standards." Believe it or not, NHTSA even issues recalls for bike racks.

When the NHTSA issues a recall, car makers (or bike rack manufacturers, one supposes) are required to fix the problem by repairing it, replacing it, or offering a refund. This is all done at no cost to the consumer. In rare instances, the manufacturer is on the hook for buying back the entire car. (Car dealerships, however, do get to bill the manufacturer; recalls tend to be seen as good news to dealers for this reason. You can read more about the sometimes contentious relationship between dealer and carmaker in Chapter 6.)

Further, automakers are required to notify you by mail in the event of a recall. You can even check whether your car is subject to an open recall (and thus a free fix) by visiting the following URL: `www.nhtsa.gov/recalls`.

TIP

The website mentioned here is a two-way street, in that it allows you to report problems about your vehicle. Your information about issues "will be added to a public NHTSA database after personally identifying information is removed." If enough complaints are received about specific issues with the same make and model of a car, the NHTSA may then investigate to see whether a recall is the best path forward.

TIP

Yes, there's an app for that. With the NHTSA's free SaferCar app, you can request to have recall information sent to your phone.

- **Vehicle safety ratings:** Now, on to the topic more germane to our discussion of EV safety, and thus to your evaluation of a vehicle before purchase.

  Before they can legally sell a vehicle in the US, every car manufacturer must submit several copies of the vehicle for crash testing, a requirement that falls under the purview of the NHTSA's New Car Assessment Program. I see these cars being shipped out of — and sometimes back into — our facilities frequently, surrounded by the grim expressions of managers and techs. One imagines a parade of soldiers passing by townsfolk before heading into battle.

  If the analogy holds, the New Car Assessment Program is the front line of that battle, where the NHTSA performs these four test crash tests:

  Frontal: I explain frontal collision tests in a bit more detail, even though there's really no need to be a crash expert to get the gist of what's going on with these kinds of tests. (Just being thorough, I guess.) Frontal test crashes simulate a head-on collision between two like-sized vehicles. The specific scenario supposes that you're headed down a two-lane road when a driver headed the opposite way starts to fall asleep and veers into your lane. In the actual tests, the NHTSA crashes the car into a fixed barrier at 35 miles per hour.

The scenario places an average-size adult male dummy in the driver seat, a small-size adult female dummy in the front passenger seat, and all dummies secured with a seat belt. Dummies are then evaluated for injuries to head, neck, chest, and femur.

- *Side barrier:* This test simulates what's colloquially known as getting T-boned — what might happen when an approaching vehicle doesn't stop at a light or a stop sign and instead slams into the driver's-side doors. The doors, one supposes, serve as the barrier between you and the front grille of the other car.
- *Side pole:* This test replicates a situation where you lose control of the vehicle, begin sliding sideways in the car, and crash into a telephone pole on the driver's side.
- *Rollover:* This is a test of what might happen if you're driving at highway speeds and suddenly come upon a sharp curve. As you try to navigate the curve, the vehicle departs the road and rolls over.

REMEMBER

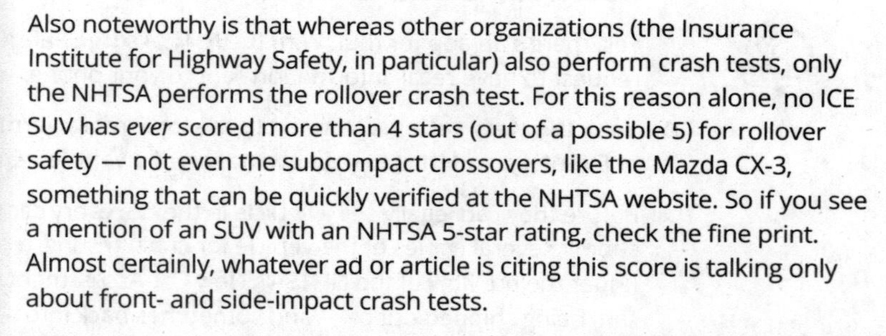

Rollovers occur in only about 1 percent of passenger vehicles, but account for *one-third* of collision-related deaths.

Also noteworthy is that whereas other organizations (the Insurance Institute for Highway Safety, in particular) also perform crash tests, only the NHTSA performs the rollover crash test. For this reason alone, no ICE SUV has *ever* scored more than 4 stars (out of a possible 5) for rollover safety — not even the subcompact crossovers, like the Mazda CX-3, something that can be quickly verified at the NHTSA website. So if you see a mention of an SUV with an NHTSA 5-star rating, check the fine print. Almost certainly, whatever ad or article is citing this score is talking only about front- and side-impact crash tests.

Unless the rollover rating is for the Tesla Model X, that is. More about that in a bit.

REMEMBER

And what about rear crash tests? The NHTSA doesn't do those, citing limited budgeting. According to its `www.nhtsa.gov/ratings` web page, they choose to "concentrate . . . ratings on front- and side-impact crashes that are responsible for the highest percentage of deaths and serious injuries."

The 5-star safety rating

Each of the aforementioned NHTSA tests is scored on a scale from 1 to 5, just like AirBnB stays or book reviews on Amazon, and just like every aspect of your personality will be in a dystopian future where Amazon, Google, and Meta all merge into one giant company. As you might guess, more stars means safer cars. (Leaving book reviews on Amazon gets you higher personality ratings from book authors. Someday.)

After you know a bit more about what a 5-star safety rating is — enough to be annoying at the next family get-together, anyway — the discussion then circles back to Tesla, to the 2012 Model S, and to the safety benchmark that was set, at least in my mind, for all electric cars to follow.

Here's what I mean.

Electric cars are, on the whole, safer vehicles. What's more, many people *expect* them to be, and this is in no small part because the Model S was such a safe car. And in 2013 the Tesla Model S *was* the entire pure-EV market. In my mind, the Model S proclaimed two statements to the general car-buying world:

- **Teslas are safe.**
- **EVs are safe (by extension).**

TIP

You can get a little more detail about the kinds of crashes performed by the NHTSA at www.nhtsa.gov/ratings. It has a couple of cool crash animations that help you visualize the kind of impact being tested, so it's definitely worth a minute of your time.

Not to recognize the impact of the Model S and Model X on consumer mindsets about electric vehicle safety would be doing both the topic and the accomplishment of the engineering teams a disservice. Had the Model S been the 2017–18 Chevy Bolt, for example — which is a very safe car, by the way; more about that later — we'd be having a very different conversation today.

X MARKS THE SPOT

Released in 2015, the Tesla Model X was *also* awarded a 5-star overall rating from the NHTSA by scoring five stars in every category. Because SUVs typically have a high center of gravity and top-heavy designs, the rollover test in particular are a tough bar to clear.

Not so with the Model X, whose battery pack (resting as it does at the bottom of the chassis) gives it a relatively low center of gravity. In fact, the NHTSA *wasn't even able* to roll the Model X over during its testing. Check out www.engadget.com/2017-06-13-tesla-model-x-earns-a-perfect-nhtsa-safety-rating.html.

Going into full Nigel Tufnel mode with the NHTSA results, Tesla began announcing that it had received "5.4 stars" based on the crash data, to which the NHTSA, playing the role of Marty DiBergi, said, "Um, dude, no." I won't unpack it all here, but the Tesla argument has some validity. However, the NHTSA safety "amps" only go to five.

But the good news is that electric vehicles have, for the most part, held to the standard set by the Model S and Model X — so much so that if you're considering a new (or used) EV today, and you're wondering which one is safest, the answer is probably yes.

I did mention that all vehicles — electric or otherwise — are tested by the NHTSA. But I also mentioned the Insurance Institute for Highway Safety (IIHS) as another testing entity, and to that I'll add another agency, the last word of which my word processor tries frantically to autocorrect: the European New Car Assessment Programme (Euro NCAP).

All do similar crash testing, and the IIHS and Euro NCAP also include tests like crash avoidance, looking over items like the driver assist features, which I discuss further in Chapter 11.

The IIHS gives out Top Safety Pick or Top Safety Pick+ awards. The Euro NCAP also awards stars on a scale of 1 to 5, similar to NHTSA.

So, with all this as background to our discussion of electric vehicle safety, which are the safest of the current crop at the time of this writing?

All cars listed here have either received 5-star NHTSA or Euro NCAP safety ratings, or have been awarded the IIHS Top Safety Pick Plus status. In no particular order:

- **Chevy Bolt EV:** The Bolt EV gets an overall 5-star rating from the NHTSA. IIHS gives the 2022 Bolt a Good rating in most categories. Both agencies suggest opting for the forward collision warning and lane departure warning features.
- **Nissan Leaf:** The latest Leafs earn a 5-star overall rating from the NHTSA (only four stars in rollover protection). The IIHS scores it as Good in most categories, and Leafs include forward collision warning and lane departure warning as standard features.
- **Toyota Prius Prime:** The Prime earns an overall 5-star rating from the NHTSA, despite four stars in the frontal and rollover protection tests. The IIHS rates the Prime as Superior for front crash protection in both vehicle-to-vehicle and vehicle-to-pedestrian collisions, and it comes with forward collision warning and crash imminent braking as standard features.
- **Mitsubishi Outlander PHEV:** It wouldn't be on this list without the overall 5-star safety rating from the NHTSA. The IIHS rates it as Superior with Optional Equipment for front crash prevention. Features like crash imminent braking and lane departure warning are included.

- **Kia EV6:** The Euro NCAP awarded the Kia EV6 a 5-star safety rating in May of 2022, just before this book went to press. It had not yet been rated by NHTSA.
- **Volvo C40 Recharge**: Volvo's company branding is centered around making safe vehicles, and so it should come as no surprise that when you combine Volvo and electric, as is the case with the C40 Recharge, you get a 5-star safety rating from Euro NCAP. It, too, has not been rated by NHTSA at press time.

I could go on. And I do, in the following list — but without quite as much detail as just listed. The cars in the following list either are IIHS Top Safety Picks or have received overall 5-star safety ratings from NHTSA or Euro NCAP (or all of the above):

- Polestar 2
- Ford Mustang Mach E
- Mercedes EQS
- Volvo XC40 Recharge
- Volkswagen W ID.4
- Volkswagen ID.5
- BMW iX
- Hyundai IONIQ 5
- Lucid Air
- Porsche Taycan
- Rivian R1T
- Audi Etron
- Nio ES8
- Renault Megane E-Tech

In fact, it's actually more of a challenge to find an electric car that *doesn't* have a 5-star safety rating than to find one that does. All these vehicles are excellent safety picks, and you can do some further reading about the topic if you're interested (although this page doesn't cover all the cars I've just mentioned): `www.iihs.org/news/detail/with-more-electric-vehicles-comes-more-proof-of-safety`.

Got a youngster or two running around and thinking of putting those youngsters in the back of an EV? Might be worth double-checking the child seat anchors on the Bolt and the Leaf. IIHS gives each a Marginal score because of difficulty of access. Note that this is not a *safety* issue, however — the anchors work as designed.

And to all of these, of course, you can add all Teslas. I've spent a good part of the chapters extolling the automaker, but it's worth mentioning that the Models S, X, 3, and Y are among the safest passenger cars that have ever existed. All Teslas include safety driver assist features such as automatic emergency braking, forward collision warning, blind spot collision warning, and lane departure avoidance.

The bottom line is that if you're shopping for an electric vehicle, you're already way ahead of the game when it comes to choosing a safe vehicle.

Knowing about the safety should put your mind at ease, or at least put it at ease for as long as it takes your <insert name of family member> to send you another news story about one of these EVs catching fire. Because as your <insert name of family member again> tells it, buying an EV is just tempting fate over the possibility of a car fire.

But is that a valid concern?

Exploring the Possibility of a Car Fire (ICE versus EV)

Okay, let's start with the obvious and then work from there:

Gasoline is a flammable liquid.

But it goes much further than that. Saying that gasoline is flammable is kind of like saying Steph Curry plays basketball. A description like that one is lacking the proper context about what a singular talent Steph Curry is. (Steph is the NBA's all-time record holder for 3-point baskets about two-thirds of the way through his likely career.) Likewise, describing gasoline as a flammable liquid leaves a *lot* of important information from that sentence.

Let's illustrate with a quick pop quiz:

Question 1: Which of these contains the most potential energy?

A. 100 sticks of dynamite

B. 1 gallon of gasoline

Yes, it's B. Gasoline actually contains about 130 dynamite sticks' worth of energy, making it kind of a miracle liquid in the same way that Steph Curry is kind of miraculous at making 3s.

TECHNICAL STUFF

Dynamite was patented in 1867 by the Swedish chemist Alfred Nobel. Yes, that Nobel. He was searching for a way to make nitroglycerin more stable. It was first marketed as Nobel's Safety Blasting Powder, which, if it isn't the same substance used during my last tooth cleaning, sounds like it should have been.

The difference between dynamite and gasoline is that dynamite releases all its energy at once, and gasoline *usually* release its energy more slowly. (Under the right conditions, gasoline too can explode.)

As you're surely aware, every ICE car carries around a tank holding 12 or 16 gallons (the energy contained in about 2,000 sticks of dynamite) of this highly volatile, explosive liquid. And during a crash, explosive, volatile liquids can, well, do what explosive flammable liquids do.

REMEMBER

If you use *TNT* and *dynamite* interchangeably, like that AC/DC song, you're doing it wrong. Dynamite is the white-powdered nitroglycerin found in sticks. TNT uses a yellow crystal in its sticks.

Obviously, carrying around gasoline is a risk that engineers have solved for already. Right? The number of car fires must be fleeting; otherwise, we'd hear more about it — right?

Let's see what the numbers have to say about these two questions.

Car fires by the numbers

In journalism, they tell you not to bury the lede, so I'll put this bottom-line conclusion at the top:

> EVs are less likely to experience a fire than their ICE counterparts.

Says who? Studies, that's who.

According to data gathered by the National Transportation Safety Board (NTSB), fully electric vehicles experience a fire 25.1 times per 100,000 sales.

That may sound like a lot — it did to me when I was first looking this stuff up — but ICE vehicles are more than 21 times more likely to have a fire, clocking in at a whopping 529 fires per 100,000 sales.

In fact, writing this section reminds me of an experience my wife had with her 2013 Ford Escape. Roughly 2 months after she bought the car, she had to take it back to the dealer because her make/model was recalled for — you guessed it! — engine fires. Almost 140,000 Escapes with a 1.6-liter engine were involved in the recall. Now that's just one example that I happened to remember because it happened to us, and I never felt unsafe in that car, but the point is that ICE vehicles and fires are no strangers.

And, if you're thinking that 2013 was, like, ten years ago (I suppose it was) and is therefore ancient history, I invite you to track down a 2021 Ram pick-up owner, who may be have a story to tell about *that* engine-fire-related recall: `www.consumerreports.org/car-recalls-defects/diesel-ram-trucks-recalled-for-fire-risk-a8454957228`.

Or, if you don't have a Ram-owner neighbor, perhaps you can hit up one of the 450,000 Kia and Hyundai owners who were told in February of 2022 to park their cars outdoors. This was part of "a long string of fire and engine failure problems that have dogged the companies for the past six years." Check out this quote and the rest of the story at `www.cbsnews.com/news/hyundai-kia-recall-vehicles-fire-risk`.

I won't belabor this point any further, as your Google searches work as well as mine. The point is that ICE fires have been happening for more than 100 years, and despite all those years to mitigate risk, there are only so many things you can engineer out of a machine whose operation requires flammable liquids.

REMEMBER

Hybrids combine both an internal combustion engine and a high-voltage battery, and therefore combine the risks of both items catching fire. A hybrid has about *twice* the chance of catching fire as an ICE vehicle.

Vehicle fires are so common, in fact, that the US Fire Administration, which is a department within FEMA (the Federal Emergency Management Administration), says that one of every seven fires — *period* — is a vehicle fire.

REMEMBER

Speaking of recalls: In 2022, there were two notable EV recalls: 82,000 Hyundai Kona were recalled because of faulty battery packs, and, in a recall that gained wider notoriety, 70,000 Chevrolet Bolts were recalled when many of them began catching fire for reasons that are, as of this writing, still under investigation. To be clear: There weren't 70,000 Bolt fires in 2022; that's just how many Bolts were affected by the recall.

So now that we've established that hurtling through space at 45 miles per hour, carrying the explosive potential of several sticks of dynamite practically in your figurative pocket perhaps isn't the safest human endeavor, let's now move to the heart of this section's issue.

Calculating the risk of an EV catching fire

EV battery packs do indeed catch on fire. But is that a common occurrence, and is this a topic that deserves a lot of your mental energy when considering a switch to electric?

To address this question, let me ask you to consider a few related ones. For example, do you own a phone? What about a laptop? Tablet? Bluetooth headphones?

You know where I'm headed with this questioning: All these items — and many more — use lithium-ion batteries. Yes, the same types of batteries powering (most) EVs are powering devices you attach to your body.

Battery fires do occasionally happen with these devices, too, but the *likelihood* of your phone or laptop catching fire is vanishingly rare. *Millions* of Li-ion products are on the market today. I'm within arm's reach of no fewer than six as I type this, and I'll sleep easy tonight with two or three of them on the charger.

In addition, electric vehicle engineers have been working hard to address the possibility of such thermal events ("thermal event" being the technical term, a euphemism really, for batteries exceeding the maximum operating temperature which often leads to a fire). Said engineers have also been mitigating the impact when batteries do catch fire. First, today's EV batteries are surrounded by a cooling layer, like how engineers design ICE radiators.

Second, in the event that a fire does occur — despite the countermeasures provided by the battery coolant — it will often be limited to a relatively small number of cells rather than the entire pack.

One caveat to all this EV fire talk

Because cars need to carry around relatively massive amounts of stored energy, all kinds of cars are prone to vehicular fire, especially after a crash.

The thing about EV fires, though, is that they are much harder to extinguish because of how they carry around that energy — namely, in the form of lithium-ion batteries. When Li-ion batteries burn, they tend to burn much hotter than the oil and gas that burn during an ICE fire, which means that these kinds of fires may need more water to fully extinguish. (At the direction of the incident commander — every emergency responder crew has a point person ultimately responsible for calling the shots — the instructions will likely be to just let the fire burn itself out.)

What's more, the battery packs can reignite hours or even days later, if their energy hasn't been fully discharged by a trained professional. This presents a risk to the location where the burned-out car is stored.

When ICE cars catch fire, emergency responders generally know what to do. ICE fires are considered Class B fires and are generally treated with both fire-retardant foam and extreme caution.

Fire personnel around the world are also now getting training on how best to deal with lithium-ion fires — I know because I help write these guides — and the guidance, for the most part, involves just waiting them out. There's nothing much to be done when battery packs go thermal other than to let them release their energy while making sure onlookers are a safe distance away.

Every EV also includes a firefighter's cutoff, a cable that, once cut, helps first responders more safely deal with crashed EVs by disconnecting the battery from the rest of the car. (See Figure 3-1.)

FIGURE 3-1: Before cutting on a crashed EV, first responders cut this switch.

electrek

This is surely a contributing factor to the newsworthiness of EV fires — they burn long enough for news crews to arrive and capture all sorts of interesting footage, and the result of an EV car fire looks like something out of a movie about an alien with a heat ray. (See Figure 3-2.)

Hey, you're still here! Because I didn't bury the lede, you got the gist of this section in the first few lines. But because you're here, I'll leave you with this bottom line on the bottom line:

FIGURE 3-2: Aftermath of either an EV fire or a carjacking by the Human Torch.

AP Photo/NTSB

All vehicles carry some risk of fire. But when it comes to fire, the data says that a pure EV is the *least* risky option you can choose.

REMEMBER

If you're standing close to a vehicle fire of any kind, don't. Get the hell away and either call first responders or wait for them to arrive. Car fires produce toxic gases and other hazards, to say nothing of the risk you face by being near an object capable of blowing up and sending shrapnel in your direction. Seriously, don't try to retrieve your fuzzy dice, don't try to douse it with your convenience store Slurpee, don't even pop the hood to see how bad it is — you can end up giving the fire oxygen, which fire loves. Just get out of there.

The leading cause of vehicle fires

I'll wrap up this discussion of vehicle fires with this reminder: According to the National Fire Protection Association (NFPA), whose job it is to gather data on these sorts of things, the leading cause of vehicle fires is this:

> Age

Not the *driver's* age, mind you — the *vehicle's*. As the NFPA's 2020 report on vehicle fires states, "[O]lder vehicles accounted for three-quarters of the highway vehicle fires caused by mechanical or electrical failures or malfunctions."

BUT EV BATTERIES *DO* CATCH FIRE OCCASIONALLY — WHY IS THAT?

There isn't a single cause of all battery thermal events, but one common reason has to do with what happens at the molecular level as the batteries are charged. Exact specifics depend on the battery, but lithium-ion batteries usually contain a metal coil and a flammable lithium-ion fluid. (See the sidebar figure.)

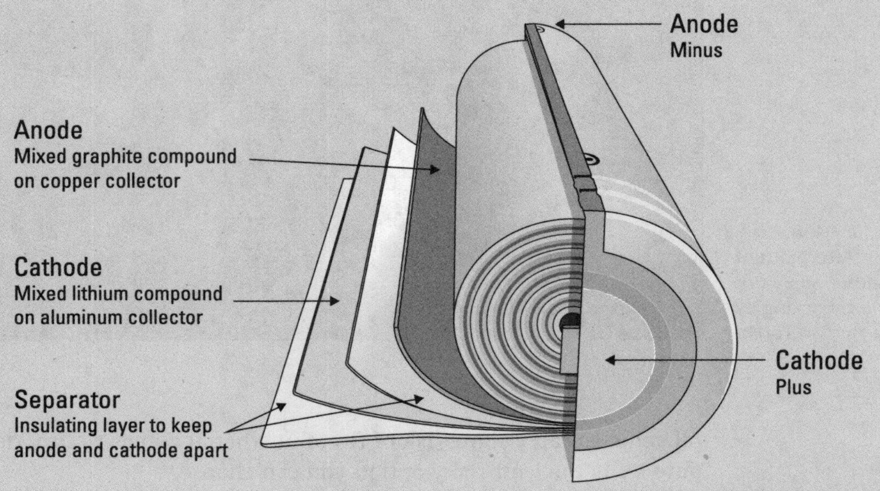

Source: Adam Bender / `https://www.adambender.info / last accessed July 11, 2022`

Essentially, tiny crystals of metal fragments float around in this fluid. The contents of the battery are under pressure, so if a metal fragment punctures the envelope separating battery components, the lithium may react with water in the air, generating high heat and sometimes producing a fire. (This is why battery-engineer types accurately refer to these as *thermal events* — not every thermal event results in a fire.)

When this happens, the battery shorts. Which means what, exactly, besides *battery bad?*

It means that all the electricity in the battery leaves as quickly as it can, because physics: The electricity will try to get to ground. Think of a battery as a tiny version of lightning in a bottle. If the lightning has a chance to escape, it will. And when it does so in a rapid, uncontrolled fashion, the result is a massive amount of heat, or at least a massive enough amount to combust many of the items typically around the battery — chairs, carpet, desks, etc.

In short, older cars and trucks are more likely to experience mechanical or electrical failures. And when you combine mechanical or electric system failure with a metal container transporting a host of flammable liquids like engine oil, transmission fluid, and — hmm, feels like I'm forgetting something.

Oh, yeah: a tank filled with gasoline.

Or, as the NFPA rather dryly puts it, "Maintenance is important throughout the vehicle's years of use."

Okay, but what if I have a pacemaker — you know, electromagnetic motors and such?

Another concern I've seen posted to various corners of the Internet has to do with not the safety of EV occupants involved in a crash but rather EV occupants who are simply *in* one.

More to the point, I've stumbled across variations of this unease, presented here in question form:

> Hey, Internet board or Comments section, I was wondering if these powerful electric motors that I'm sitting just a few feet from pose some kind of danger if I'm wearing a pacemaker or defibrillator or any other device that sets off airport scanners?
>
> Oh, and as long as my keyboard is working, what about electromagnetic radiation? Is that something I should be worried about before buckling into an EV? I mean, I'm not the tinfoil hat type or anything, but seriously: Do I need a tinfoil hat?

Fortunately, your author has an app that can help craft the perfect response to these concerns. The app is called . . . Science!

First off, know that you and every soft, squishy sack of mostly water walking around on two legs — that would be a human — is being absolutely *bombarded* by electromagnetic radiation, thanks to the sun. Sunlight, which most animals can see some wavelengths of, is part of the electromagnetic radiation spectrum. Infrared light is also part of that spectrum, but is invisible to the human eye. (It can be seen with human-created tools, however, which is kind of badass. Thanks, science!) Microwaves and radio waves are also part of the same spectrum. So, whether you're heating up your coffee at 2 P.M. or listening to an afternoon baseball game, electromagnetic radiation is part of your daily existence — you have some control over the microwaves, but not the radio ones. Changing stations on a radio is proof that the waves are being sent along multiple frequencies 24 hours a day, 7 days a week.

In fact, you can't escape electromagnetic radiation even when you're sitting in your house, in the dark, at night. Yes, you're even absorbing electromagnetic waves while wearing a tinfoil hat.

Now, in fairness, certain wavelengths of electromagnetic radiation aren't so great for the human body. At the higher end of the EM spectrum — the wavelengths here are shorter and, generally speaking, contain more energy — are ultraviolet waves, which cause sunburn, x-rays, which can be quite harmful in large, prolonged amounts, and gamma rays, which are common in space, but on Earth are associated only with nuclear energy.

REMEMBER

The electromagnetic radiation worth worrying about, sorta, is *ionizing* radiation. Ionizing radiation means that the light wave contains enough energy to strip away electrons from atoms when the two interact. Examples of ionizing radiation are ultraviolet, x-ray, and gamma ray. But it's a matter of degree. I can take *some* UV radiation, but as a pasty redhead, if it's going to be more than ten minutes, I try to remember the sunscreen. I've had some x-rays on my knee and during dental check-ups, but it's not something I'd want as a daily occurrence. Fortunately, Earth's atmosphere protects us humans from most of the face-melting gamma rays that ricochet around outer space after stars go supernova. So, yeah — at the rate we're going, we may need a tinfoil hat for the *planet*.

So-o-o, are you going to get a sunburn, or worse, riding in your EV? Is a Chevy Bolt delivering literal bolts of invisible radiation while driving?

In a word, no.

Yes, EV motors do rely on electromagnetic radiation — the charge passing through the coils generates the rotating magnetic field that causes the motor to turn.

But no, the kind of electromagnetic radiation produced by electric motors isn't the kind that causes harm to human tissue. The radiation is *non-ionizing*. (Please tell me that you read the Remember paragraph, a couple of paragraphs up.) What's more, the amount of radiation produced by the EV motor is just a fraction more than the radiation produced by an ICE car. That's right: ICE vehicles *also* produce electromagnetic radiation!

If you're *really* wanting to worry about radiation while driving, you're better off focusing on the window, not on the motor.

In sum, though it might be super gnarly for your son or daughter during a bring-your-dad-to school-day, the fact is that driving an EV won't pump you full of gamma radiation, allowing you to *Hulk*-out at will. (Will Marvel sue us for using the words *Hulk*-out? Only one way to find out.)

TECHNICAL STUFF

If you want to nerd out about the topic of radiation, do some reading about the inverse square law: Essentially, the further away you are from a source of radiation, the safer you are, and it's not a linear progression. In other words, standing two feet away from a source of radiation doesn't mean that you're getting half as much radiation as if you were standing one foot away.

TL;DR: Unless your passenger is, in fact, the *Hulk,* and you make them angry, from an electromagnetic standpoint, you're perfectly safe in an EV.

That's true today and will continue to be so in the future.

A Car That Gets Safer Over Time

A kind of double meaning is at work in this section's heading. The first meaning is that the EV is a *type* of car . . . that gets safer over time. The second meaning implies that you can expect *your* EV . . . to be a car that gets safer over time.

So let's explore each of these twin implications in more detail.

EVs will get safer over time

As the calendar months continue to advance, so too does the safety engineering being done on electric vehicles. What's more, cars in general are getting safer over time — from safety glass to seat belts to advanced driver assistance features — and so it stands to reason that the safety advances in the coming years will extend to electrics, if only because nothing else will be manufactured by several prominent automakers.

What might that safer EV future look like?

As you've seen, most vehicle safety engineering before the early 2000s (give-or-take) focused primarily on the physical — keeping occupants from being injured or killed in a crash with design elements like front-impact pillars, airbags, seat belts, safety cells, and even better tires.

That will certainly continue as EVs are brought to market, and some really smart people are working on this problem all the time. They're trying to make the car chassis safer, utilizing battery tech that's less prone to thermal runaway, and publishing PhD research on how to make the lithium-ion batteries now in use even safer. (See the research in this article for an example: `https://me.engr.uconn.edu/blog/2021/09/16/mechanical-safety-of-lithium-ion-batteries-for-electric-vehicle-applications`.) In short, the march of progress in the world of atoms will continue apace.

Cars made in the 2020s and beyond, however, try to reframe safety engineering in terms of not getting in a crash in the first place. How?

By mostly focusing on safety as expressed by the car's bits — all the 1s and 0s that handle much of the command-and-control of a modern car.

And bits, of course, means software.

In the pursuit of not even *needing* better crumple zones, airbags, and the like, today's cars use a combination of sensors, cameras, radar, and raw computing power that was more associated with supercomputers than vehicle not so long ago. Today's cars and trucks use processors that are equal to those powering the fastest gaming consoles. In most cases, these chips come from the very same manufacturers — AMD, Intel, Nvidia, and the like.

Taking this perhaps one step further, Tesla showcased its own Dojo supercomputer in late 2021, built entirely in-house. Tesla claims that Dojo is capable of processing camera data four times faster than any computer on the planet, with the idea that Dojo-trained artificial intelligence (AI) is then used to improve self-driving in more frequent increments.

With these advances in sensors, cameras, and the software used to act on that input, you can soon expect to see electric vehicles with even better safety performance than the ones on the road today. Specifically, you can keep an eye out for enhancements like these:

» **Computer vision:** This term describes a combination of human and computer vision — think the kind of HUDs, or heads-up displays, used by pilots in fighter jets, except even better. Computer vision works by casting a 3D holographic animation into the driver's field of view. (See Figure 3-3.) The display might highlight a turn from the navigation without having to take your eyes off the road, or it might alert drivers about issues like stopped cars ahead or pedestrians waiting to cross at a stoplight.

FIGURE 3-3: Computer vision.

Brian Normile

Computer vision is being worked on by firms like GM, Panasonic, and Continental. It uses a heads-up display that can track a driver's eye movements to ensure that vital information is projected in their line of sight, no matter where they look or how far they sit from the steering wheel.

- **Heat death prevention:** The National Safety Council reports that an average of 38 children ages 15 and younger die of heatstroke every year while in vehicles. (I warned you that reading about car safety was no fun.)

  A combination of sensors, cameras, and software can, hopefully, prevent this tragedy altogether someday soon. The idea is straightforward: Either an internal camera or sensor would detect that a child (or pet) is in the vehicle, which would, in turn, set off alerts and warnings. Additionally, software could turn on the car and start up the air conditioning.

  Something close to what I've just described is already in use in select cars like the Hyundai Genesis, which isn't an EV, and in Tesla vehicles using something called Dog mode. EVs also have the advantage that the software doesn't have to determine whether the car is inside before turning it on and starting the air conditioner, because there's no exhaust to worry about.

- **Dooring prevention:** If you ride a bike in an urban environment, you're probably familiar with the gnawing dread of being doored when passing a parked car. *Dooring* happens whenever someone opens a door directly into the path of a cyclist, and further happens because city engineers tend to paint bike gutters next to lines of parallel-parked cars. Dooring sounds like it would result only in damage to a car door and a skinned knee, but it accounts for about 20 percent of cycling crashes in cities, and the results can be fatal.

Automakers like Audi, Mercedes-Benz, and Lexus are already using sensors to detect oncoming bicycle/vehicle traffic from behind. You may have seen recent ads from Hyundai and Kia promoting the Safe Exit Assist system, which engages the child safety locks to prevent rear doors from opening when approaching traffic is detected.

The takeaway here is that these collision avoidance technologies should be coming to an EV near you sometime very soon.

- **Adaptive driving beam technology:** If you've ever been blinded by approaching headlights when driving along a two-lane highway, you know that the feeling of discomfort isn't just in your retina — it's also in the pit of your stomach. Adaptive driving beam (ADB) headlights — sometimes marketed as smart headlights — do everything modern headlights do from a visibility standpoint while including technology that prevents headlights from frying the eyeballs of oncoming drivers. Because that's just not nice.

  Some recent ADB advances facilitate the headlight array spotlighting pedestrians or cyclists, or help project virtual lane lines on snow-covered roads. So, if you pull your new EV up to your garage and see something like the dark spot shown in Figure 3-4, your headlights don't need service. The dark spot is there by design.

REMEMBER

When talking about these safety advances, even if your EV doesn't have them now, it very well could. And better still: It might be able to add these safety features without you having to make a trip to the dealer.

FIGURE 3-4: Working as designed — approaching drivers will thank you.

Porsche Cars North America, Inc.

Your EV will get safer over time

Here's the scenario: Let's say you really want one or two of the features mentioned in the preceding section, but the current model you're shopping for doesn't yet have that feature. What is an EV buyer to do?

My take is that, if you have the financial means to do so, buy the EV today. You can start enjoying all the other advantages of an EV, and the safety advances you're pining for might be just one software update away.

Why? Because many, if not all, of the safety features of the future don't rely on available hardware. All cars come with locks, for example. Cameras exist and have been added safety features for years. It's the same with sensors, radar, and so on. The difference is in how the car's "brain" uses this data input and makes subsequent decisions. And making better decisions — what to do with car locks as data is collected about a possible cyclist approaching from a human blind spot — is a function of the software, not of the door locks or cameras.

I first tackled the subject of over-the-air (OTA) updates to Chapter 2, so please refer back to that chapter for a full breakdown if you haven't already read through it.

The way OTA updates relate to safety is that most manufacturers are including these software improvements as a standard feature, which means that in terms of safety, you should receive the latest updates automatically — a change to how your headlights work for example, improvements to battery pack management, or code that might send an alert to your phone (or just call you) when the interior-facing cameras detect animals or children left behind in car seats.

With almost any late-model EV, *your* particular car can get even safer than it was on the day it was brand new, and all this can happen from the comfort of your garage, and without you having to remember to request the update.

Your EV will get safer over time

2 Buying Wisely

IN THIS PART . . .

Explore how much an electric vehicle *really* costs you

Compare EV costs with the costs of a gas-powered vehicle

Determine which EV option works best for you and your needs

Sign the papers and take the keys

» **Comparing gas and electric**

» **Paying for maintenance**

» **Keeping other car ownership costs in mind**

» **Being mindful of the bottom line**

Chapter 4

Consider the Cost (of an Electric Vehicle)

Hmm. How to introduce the chapter on the accounting side of an electric vehicle? I could say something like, "You'll have to pay for the car, you know," but that would be stupid.

If I were not paying homage to a favorite author with the chapter title, I'd call this chapter something like "Exploring the Price Parity Equation of Gasoline Versus Electric," but that's a mouthful of a chapter title (to say the least), so instead let's just shorten it to this:

Price Parity

It's a word that enters many an electric car chat (and just stop with your objection, "But that's two words" when the *one* word is *parity*), and when the word is rephrased as a question, what you get is something like this:

When will electric car prices be the same as internal combustion cars?

As you will soon see in this chapter, my answer to that question is *now*. Why I've come to that conclusion, then, is what we explore throughout. We'll start by comparing the yearly cost of ICE versus EV ownership, then start specifically measuring the cost of fuel, and finally look at other items that are part of the total cost of ownership equation. When we're done, I think you'll have reached the same conclusion.

Let's dig in.

TIP

In case you're curious, the chapter title owes its existence to the late author David Foster Wallace (DFW), who wrote the essay "Consider the Lobster," from a book of the same name. The essay is about a summer lobster festival in Maine. It's how I learned that it was once considered inhumane to feed lobsters to prisoners and, like everything Wallace writes, it is brilliant, funny, revelatory, empathetic; it's ultimately much more about what it is to be human than what it is to attend a lobster festival. It also has more than a little subtext about what humans have done and are still doing to the natural world.

As you can see, I highly recommend the essay. Lastly, I'll add that in order for this note to be a proper DFW homage, it would have appeared in a footnote. Alas.

The Price of an Automobile: An Analysis Using Numbers

Owning an automobile isn't cheap, by any stretch of the imagination.

And before you say, "Do you actually get paid to type things that everyone knows into sentence form — is that really a thing?" let's first deal with the instance where people receive a car for free. It does happen. (Oh, and the answer to the question was, *Yes.*)

Even if you have a parent or another relative who wraps up a set of old car keys and gives them to you as a birthday or graduation present, you still face several other monthly and yearly costs associated with owning that car.

Specifically, after your name is on the title and assuming you don't have an account that appears on Rich Kids of Instagram, you'll find yourself sweating over costs like these:

- Insurance
- Registration
- Maintenance
- Fuel

How much does the free car cost? According to the US Department of Labor, it's just over $6,000 per year. Here's how that amount breaks down:

- Gas and oil: $2,100
- Other vehicle-related costs (taxes, insurance, maintenance): $4,254

Six thousand dollars. After tax. For your free car. And let's put a pin in that $2,100 figure for the time being.

Now let's extend that same calculation to the amount the typical American spends on the car itself. That figure, again according to the US Dept of Labor, is $4,400.

What do you get, then, for the average automobile cost for the car-driving public?

Ten thousand dollars.

Again, this is *after-tax* money spent on your personal transportation. The exact average, by the way, is actually a bit higher than that, checking in at $10,742 at the time of this writing, but we'll round down for the sake of argument.

Now let's translate those dollars into labor.

Say that you work a job that pays you $20 per hour. That's *about* what Amazon pays most of its warehouse staff. How many weeks of your life, then, do you have to work every year in an Amazon warehouse just for the privilege of driving to that warehouse, assuming that you buy an average-priced US (and thus ICE) automobile?

It's 16 weeks.

In case you're wondering, a $4,400 per year annual car cost nets out a $365 per month car payment. For further comparison, financing $20,000 over 60 months at 4.5 percent gets you about to that point when factoring in typical registration and sales tax expenses.

The back-of-napkin math works out like this:

$12,500 pretax earnings ÷ $20 per hour ÷ 40 hours per week

That works out to 16 weeks (or 15.6, to be exact).

The bottom line is cars demand a lot of our money, a lot of our labor. The bottom line is that for a worker making $20 per hour, even a "free car" could cost you multiple *months* of going to work every day. Maybe you should stay in the good graces of that relative instead.

Gas versus Electric: A Comparison Using Numbers

The reason I'm walking you through a discussion of a typical car's annual cost is it's a pretty good starting point for the analysis of price parity. That is, when considering price parity, $10,000 is a fairly good starting point from which to do this mental estimation. You're going to spend $10,000 per year on a car. The fact that it's an average means that many people will spend more than that. Okay, fine — you can afford it.

The question is, though: Do you want to spend it on the car or on the fuel and maintenance of the car?

Because, again, the direct cost of a typical ICE car is $4,400 per year. The bulk of the $10,000 spent to own a car is spent on items that *aren't* the car itself. In fact, if you could spend all of the $10,000 of annual car expenditures on just the car, and if the maintenance and fuel were free, that would allow for a car payment of

$833 per month

Wow! That's kind of a big car payment.

Where all this is headed, of course, is that seeking out ways to drive down fuel-and-maintenance costs is going to allow you to get yourself more car for your $10,000 yearly transportation spend.

As it happens, $833 per month with zero down, financed over 72 months at 4.5 percent, will get you in a car that costs $52,000. And $52,000 will get you behind the wheel of quite a few cars, electric or otherwise.

If only there were a way to spend almost nothing on fuel-and-maintenance costs, that is. If only such a type of automobile existed.

Hmm. If only . . .

Oh. Wait!!

So, about that price parity thing

As we've just seen, for many Americans (I'm saying Americans because I'm using the US numbers here to illustrate), a lot of work is required to facilitate driving to work.

So, before we dive into the numbers of why going electric will likely help you drive down the overall cost of car ownership, I'll note that the best advice I can pass along about car ownership from a *financial* standpoint is not to own a car in the first place.

In fact, I'll have more to say about this matter in Chapter 13. For now, I just want to make sure you've considered other transportation options first — options such as these:

» **Public transportation:** Trains, buses, subways. Granted, the options can be either amazing (the Netherlands), dreadful (almost anywhere in the US), or nonexistent (almost anywhere in the suburban wastelands of the US). A quick anecdote about this: When my family moved from Kansas City (one of the worst places on earth, in terms of mass transit) to San Francisco, I started taking the Bay Area Rapid Transit (BART) to and from work every day. We sold one of our cars. We had more spending money every month as a result. And you know what else? I was happier because I didn't spend time in traffic every morning and afternoon.

  What's more, cities, counties, and provinces are electrifying their public transit vehicles in order to meet mandated or voluntary emissions standards. (The bus you ride around Banff National Park in Alberta, Canada might not have a tailpipe.)

» **Biking:** Again, whether or not this is an option is mostly a factor of whether your city's leaders view cities as places for people (again, I'm thinking of The Netherlands) or for cars (I'm thinking of the US in places like Overland Park, Kansas, and Fremont, California). Now that that's out of the way, biking to work has so-o-o many upsides to physical and mental health, and results in your being able to afford more date nights with your spouse, which in most instances is a very good thing.

  Riding an electric bike lets you go further, faster, easily covering most commutes to work or school. Plus it saves you massive piles of money when compared to the expense of a car and improves your health to boot. There isn't much *not* to like about going by bike.

  As mentioned, I have a whole thing on this in Chapter 13, and it might just be the most impactful section for many who read this book.

» **Carpools:** I get it: People need cars. They simply can't bike or take public transportation, given their current circumstances, which in turn may have been influenced heavily by parents or employers or other factors (like living a long way from public transit) that are beyond their reasonable control.

  Carpooling is an option that saves you wear-and-tear on your car, which saves you money, and as a side benefit, you're likely to develop closer friendships as a result. Quantity time is quality time.

AND YOU WONDER WHY SAAB SWITCHED TO BUILDING ANTI-TANK WEAPONS?

While we're on the subject of car affordability: The most expensive car I ever owned, in terms of price versus maintenance, and also in terms of price versus stress level, was a 1987 Saab 9000 that I picked up for $3,000 and paid for with a check that bounced.

One year later, after replacing practically every single part of the engine, work that was done by the city's lone mechanic certified to repair those Swedish cars that wasn't the dealership, I had spent nearly three times that amount. The cheap transportation didn't exactly end up being that cheap, after all. RIP, Saab Auto, and good riddance. (And yes, I remedied the bounced check; it wasn't on purpose.

Were you to shop for a Saab automobile on the Internet today, you might be surprised to learn what the Stockholm-based manufacturer now sells. I guess I gave it away with the sidebar title, but yes: Saab no longer makes cars like the 900 or the 9000, instead focusing its collective brainpower on weapons of war — sorry, "advancing the basic human need to feel safe." In Silicon Valley, you hear lots of companies make lots of, um, *aspirational* claims about what they really do, but that is some weapons-grade yogabab-ble right there, Saab.

Sorry. Writing about my old 9000 is a bit triggering.

Estimating the cost of gas for ICEs

So now let's return to the average fuel cost of an ICE vehicle, which again is $2,100, according to the 2019 US Bureau of Labor Statistics.

This is a good, round number for estimation purposes, but it's obviously an esti-mate that's a bit dated. As you consider purchasing your first EV, you might want to have more recent numbers, for comparison's sake. The gas prices of today aren't what they were in 2019: They might be higher by the time you're reading this chapter, or they might be lower — my parents threw out my Magic 8-Ball when they (gleefully) turned my bedroom into an office.

In any event, calculating an ICE car's annual fuel costs is a factor of these three numbers:

» Miles driven in a year
» Vehicle's fuel efficiency
» Cost of gas

TECHNICAL STUFF

According to the EPA, the average 2022 vehicle gets 27 MPG.

So let's suppose that you drive 15,000 miles a year (which is how annual fuel spending is calculated by the EPA) and you have a 4-cylinder sedan, like the BMW 530i or the Toyota Camry XLE AWD, both of which get a combined 28 MPG. Let's further suppose that gas costs $4 per gallon (quite a bit *less* than current prices at the pump here in Cali).

TIP

To find out your family's annual fuel spend on that sedan, you can go to a website like `www.calculator.net/fuel-cost-calculator.html` and enter the numbers manually. Alternatively, you can go to the EPA website at `www.fueleconomy.gov` and just select the make and model of the car(s) in question. Again, you have many sites to choose from that are only a Google search away. The EPA in particular has a fantastic site because, generally speaking, the government is really, really good at stuff like this. (See Figure 4-1.)

What do we get when we run the numbers as this chapter goes to press?

> For the BMW, you'll spend $3,100 per year in fuel.
>
> For the Toyota Camry, you'll spend $2,400 per year.

Incidentally, the Camry comes out ahead in fuel costs at the EPA website because it assumes premium gasoline for the beamer, which of course will cost more per gallon than the Camry's regular gas. I had to stare at it for a little while, too.

(Oh, and a quick a user interface aside: the eye tends drawn to the "save or spend" row of the EPA website's comparison table, but these are the savings/spending over 5 years. So if you spend $1,500 more than the average 2022 vehicle over five years, as is the case with the BMW that's only $300 per year, or $25 per month. Not really worth fretting over, especially for folks in the market for a BMW or similar.) Fortunately, several resources can help you perform these same estimates for EVs, providing a useful barometer against which to measure your potential cost savings when going electric.

Estimating the cost of electrons for EVs

Okay, now let's turn to electrons.

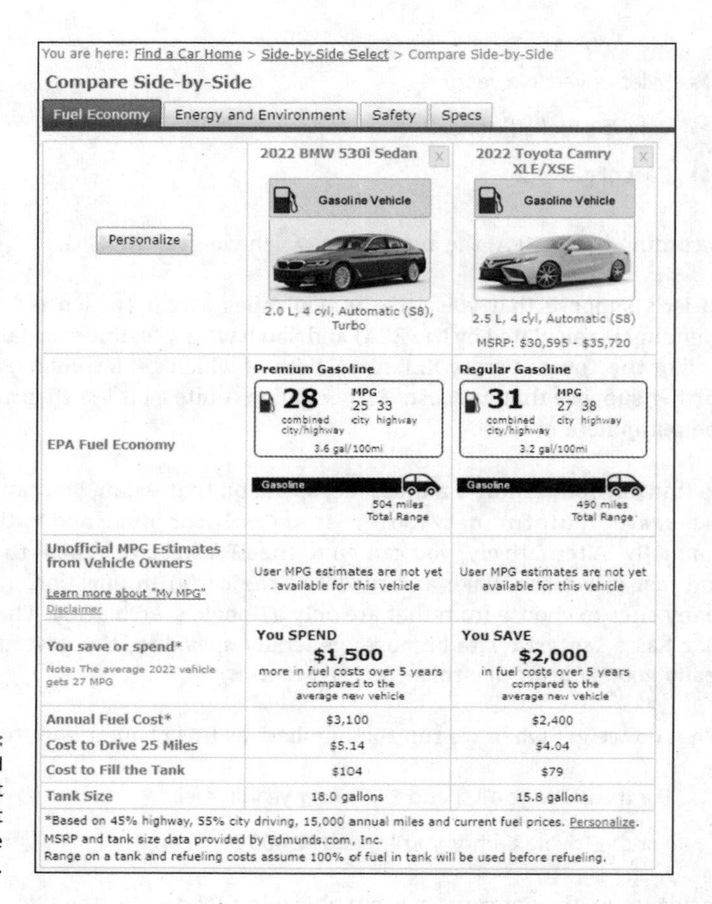

FIGURE 4-1: Why you should thank the next government employee you see.

To do so, you might not even have to leave the website where you're estimating your potential gasoline costs. For example, from the same EPA website used to forecast ICE fuel spending, you can easily throw two other cars into the mix. I recommend these:

- Tesla Model 3 Long Range AWD
- Hyundai Ioniq 5 AWD Long Range

What do we get when estimating the propulsion costs of these cars?

- For the Tesla, you'll spend $500 in fuel.
- For the Hyundai Ioniq, you'll spend $650.

Figure 4-2 shows the results in black-and-white.

The implications of this are profound.

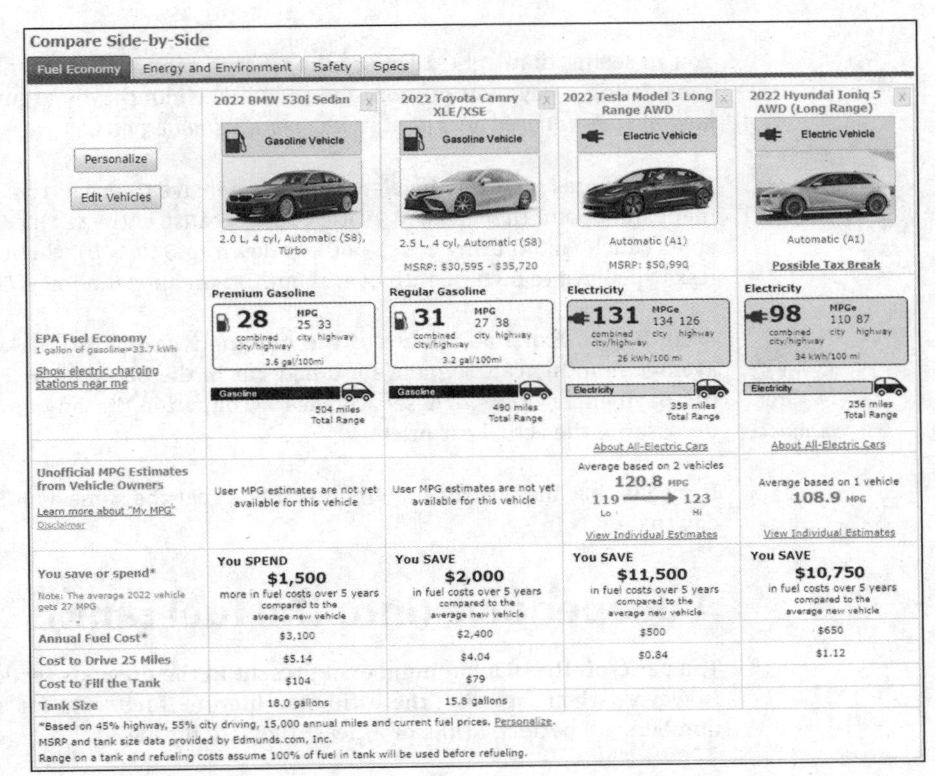

Compare Side-by-Side

Fuel Economy | Energy and Environment | Safety | Specs

	2022 BMW 530i Sedan	2022 Toyota Camry XLE/XSE	2022 Tesla Model 3 Long Range AWD	2022 Hyundai Ioniq 5 AWD (Long Range)
Personalize Edit Vehicles	Gasoline Vehicle 2.0 L, 4 cyl, Automatic (S8), Turbo	Gasoline Vehicle 2.5 L, 4 cyl, Automatic (S8) MSRP: $30,595 - $35,720	Electric Vehicle Automatic (A1) MSRP: $50,990	Electric Vehicle Automatic (A1) Possible Tax Break
EPA Fuel Economy 1 gallon of gasoline=33.7 kWh Show electric charging stations near me	Premium Gasoline 28 MPG combined city/highway 25 city 33 highway 3.6 gal/100mi Gasoline 504 miles Total Range	Regular Gasoline 31 MPG combined city/highway 27 city 38 highway 3.2 gal/100mi Gasoline 490 miles Total Range	Electricity 131 MPGe combined city/highway 134 city 126 highway 26 kWh/100 mi Electricity 358 miles Total Range About All-Electric Cars	Electricity 98 MPGe combined city/highway 110 city 87 highway 34 kWh/100 mi Electricity 256 miles Total Range About All-Electric Cars
Unofficial MPG Estimates from Vehicle Owners Learn more about "My MPG" Disclaimer	User MPG estimates are not yet available for this vehicle	User MPG estimates are not yet available for this vehicle	Average based on 2 vehicles 120.8 MPG 119 Lo → 123 Hi View Individual Estimates	Average based on 1 vehicle 108.9 MPG View Individual Estimates
You save or spend* Note: The average 2022 vehicle gets 27 MPG	You SPEND $1,500 more in fuel costs over 5 years compared to the average new vehicle	You SAVE $2,000 in fuel costs over 5 years compared to the average new vehicle	You SAVE $11,500 in fuel costs over 5 years compared to the average new vehicle	You SAVE $10,750 in fuel costs over 5 years compared to the average new vehicle
Annual Fuel Cost*	$3,100	$2,400	$500	$650
Cost to Drive 25 Miles	$5.14	$4.04	$0.84	$1.12
Cost to Fill the Tank	$104	$79		
Tank Size	18.0 gallons	15.8 gallons		

*Based on 45% highway, 55% city driving, 15,000 annual miles and current fuel prices. Personalize.
MSRP and tank size data provided by Edmunds.com, Inc.
Range on a tank and refueling costs assume 100% of fuel in tank will be used before refueling.

FIGURE 4-2: Using electrons as fuel tends to move decimal places to the right.

Price parity: The wait is over

Let's take a look at the $35,000 Camry. When the loan is financed for 60 months at 4.5 percent, the monthly payment on a Camry comes out to about $650 per month. Now add in about $185 per month in fuel, and the total cost to operate the car, exclusive of maintenance costs, is

$835 per month

For the $57,000 BMW, we're looking at about $1,275 per month to own and operate, exclusive. (I describe the maintenance aspect in another chapter — specifically, Chapter 9.)

Both the Hyundai Ioniq 5 and the Tesla Model 3 cost $47,000 in early 2022 (and this is after two recent price hikes from the latter). Paying the amount of $47,000 over 60 months at 4.5 percent is

$875 per month

You're seeing that right: Give or take what is essentially a rounding error when considering a payment of more than $800 per month, my argument is that *EVs have already achieved price parity with their ICE counterparts.*

In other words, a Tesla Model 3s is less expensive to drive from a monthly payment standpoint than a BMW 5 series. And because of the tax incentive now available (which would bring a $47,000 car down to $39,000), the Ioniq 5 in 2022 is less expensive to own, month over month, than almost all new Toyota Camrys.

TECHNICAL STUFF

According to Experian's State of the Automotive Finance Market, Q3 2021, the average monthly car payment for a new car in the US is $609. The average used car payment is $465. These averages account for the amounts financed, not necessarily the actual car price.

It means that the answer to "When will EVs cost the same as ICE cars?" is "Any day that ends in -y."

A deeper dive (into the fuel tank)

The back-of-the-napkin numbers I present in the previous section are just that. However, when factoring the estimated lifetime fuel costs for a vehicle, round numbers are perfectly fine, or at least they are for me.

Why? Because over five years of ownership (the average in the US is actually almost seven now), will it really matter whether I save $8,000 versus a gasoline car or $12,000 if I anticipate either rising gasoline prices or an even longer term of ownership? Or $6,000 if EV demand results in oil prices crashing — a not-unlikely scenario?

There are too many variables, almost none of which I can project with any certainty, and none of which answers the most important questions: "Will I enjoy driving the thing every day? Is it quiet? Comfortable? Safe? Will it sound like a concert hall when I want to sing along to the *Hamilton* soundtrack? Does it convey a status commensurate with the perception I have of myself?"

Until I can time-travel, there are only so many items I'll keep in my mental RAM, and fuel costs get a little of that space, but not much. Car ownership will be an expensive proposition, either way. (My preference, in case you're wondering, is toward self-driving rides, which is yet another scenario I explore in Chapter 11.)

All of that said, if you really *want* to dial in your electric fuel estimates as they exist today, you can head over to a site like `chargeevc.org`. There you'll find this handy resource: `www.chargevc.org/ev-calculator`.

From here, you can plug in your daily driving estimates and adjust variables like how efficient your EV is in terms of miles per kWh — this will be a pre-ownership estimate, and 3.5 miles per kWh is a close-enough estimate for most purposes; you should be able to get this info from the car directly after it starts logging miles. The calculator then spits out what you can expect to spend annually on electrons.

However: Suppose that you've chosen to live in an exurb of Houston, Denver, Phoenix, or Washington, DC. Firstly, ugh — move away and you'll be much happier, but I digress. Secondly, let's say you commute 50 miles each day to work and on errands and you put in 80 miles per day on the weekends carting Johnny to his various "select" baseball tournaments or driving you to your tee time. (Yes, I'm making some assumptions here about why you put that many miles on your car.)

Though someone who has chosen to put so many miles on a car every year is likely a) American and b) doesn't need the savings, let's do the math for the sake of argument. (See Figure 4-3.)

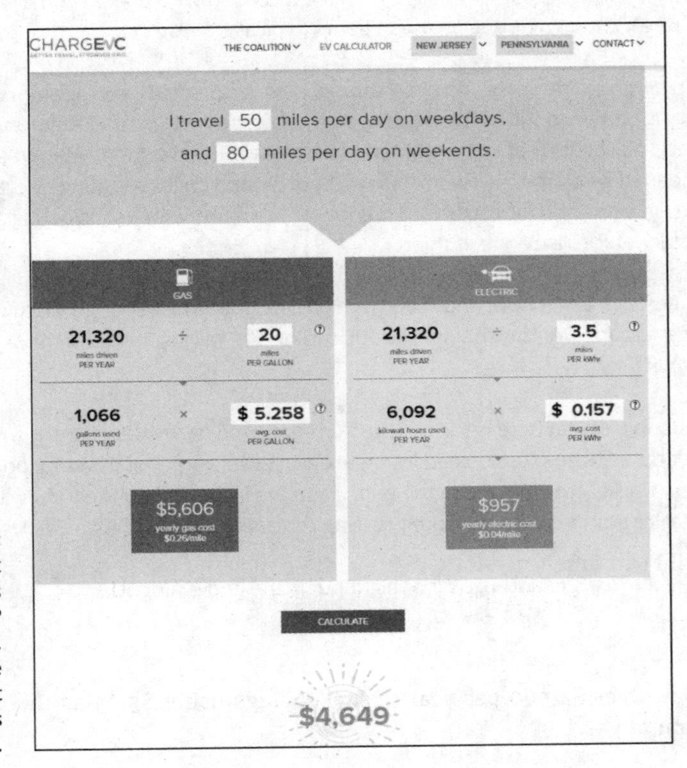

FIGURE 4-3: Saving the planet for animals and children? Great, I suppose. I thought we were talking about *other* savings here.

THE GRIDS ARE ALRIGHT

During my many EV travels, I've noticed that a nonzero number of folks considering EVs (or dismissing them) seem to do a lot of worrying about "the grid."

As in there's no way the grid could handle it if everyone were to go out and get an EV — the poor grid!

Strangely enough, however, I don't hear any of these same people worrying about how the grid handles all those toasters running at 7 A.M. on a weekday. Or refrigerators in every house and restaurant in the world. Or air conditioning on a Phoenix summer afternoon. OK, maybe they do talk about this last point from time to time, but the point is that utilities know with great precision what their current needs are, and what trends they need to plan for. As a result, power companies almost always have more generating capacity than they need — in fact, they're regulated in order to make this so.

Besides, the fact remains that, for the most part, EVs soak up electricity when there's plenty of excess capacity (at night or in the morning while parked at the workplace) and are rarely charged during peak use hours (typically, 5–9 P.M.).

As it is with airline seats and hotel rooms, excess capacity is worthless — utilities have every incentive to sell all the capacity they can generate. So, until I start seeing widespread blackouts that are attributed to electric vehicle charging — which to-date has happened zero times — I won't burn a lot of mental cycles sweating "the grid." To be sure, an aging grid infrastructure has lots of problems worth considering, but charging electric vehicles isn't one of them.

Besides, EVs aren't now, and likely never will be, made fast enough to truly strain electrical generating capacity. It's not like 100 million EVs will just fall from the sky next Wednesday.

(A possible caveat here is the use medium-duty and heavy-duty electric trucks. This isn't a sidebar of absolutes; instead it's meant as reassurance that EV adoption *in and of itself* is very unlikely to overwhelm the grid. If you live in Texas, you likely know that an occasional ice storm is a much more pressing concern for the grid than EVs.)

Or, as the British band The Who might put it, the grid is alright.

Boom. Over $4,500 per year in fuel savings using $5.25 as the going rate for a gallon of gas.

The larger point is this: The more miles you log in a week of driving, the more your fuel savings adds up when logging those miles in an EV. The range anxiety most people feel *should* be of the kind that impacts their wallet, not the distance the EV travels.

The French-Canadian connection

This won't be the last time I apologize for this, but I must beg forgiveness for the American-centric presentation of the fuel savings case here.

I do realize that there are places other than the US — Canada, for example. I know this because I lived in Canada for over a year. (And I hope to do so again. Either there or Amsterdam. Or New Zealand. Maybe Norway. Oh, and it was Vancouver, if you must know.)

In any event, gas prices in Ottawa (where French is an official language — hence, this section's heading) are $1.50 as I write this chapter, which may sound inexpensive until you realize that Canada, like most everywhere else, uses the metric system. That's the price *per liter*. Or *litre*, perhaps. What*ever*, Canada.

The amount $1.50 per liter works out to about $5.70 per gallon. In fairness, that's in Canadian dollars, so I think that in turn works out to 83 cents per US gallon, although some astute reader will have to check my math. I don't have a calculator handy, and this is my second jab through the Peace Arch separating our neighbors to the north in the same number of paragraphs! (Peace Arch Park is a lovely spot between the customs stops at Interstate 5 and Highway 99.)

I won't belabor the exact calculations because, at the end of the day, it's forecasting, which, as I've mentioned before, is hardly an exact science. The bottom line, though, is that Canadian EV fuel savings are even more pronounced than in the US: a savings of almost $10,000 over five years, assuming that gas prices remain at current levels.

Here are some other places where cars exist and gas is sold to fuel those cars, in US dollars per gallon prices because I want to humor my US audience:

- Norway: $9.00
- France: $7.10
- The Netherlands: $6.48
- Croatia: $6.62
- Japan: $4.97
- Hong Kong: $10.87
- Saudi Arabia: $2.35

So, yeah. No wonder Norway is buying so many EVs. And if you absolutely *must* drive in Hong Kong, it probably pays to get an EV.

Of course, fuel isn't the only routine and generally recurring cost associated with operating a car. The next few sections look at the other costs.

EVs and Maintenance: An Estimation Using Numbers

Cars are complex machines, and complex machines eventually break.

Fixing what goes wrong in a complex machine will add cost to ownership — period. Yes, you may see a few exceptions here and there (a very short-term lease/ownership run or a lease where all maintenance is included, although I've never seen one that includes items like tires), but there is no avoiding maintenance costs. The question for most car owners becomes, "How can you minimize those maintenance costs?"

The average cost of an automotive repair is around $600; check out the numbers here: `www.aaa.com/autorepair/articles/planning-for-auto-maintenance-and-repair-costs`. And thus one of the best ways to avoid maintenance costs is not to have maintenance costs in the first place.

Which brings us, as always, to electric.

Electric vehicles have fewer moving parts than ICE vehicles — that's just the way it is. And fewer moving parts means fewer points of failure. *Consumer Reports* recently conducted research comparing the cost of maintenance for electric vehicles and gasoline cars, and electric came out far ahead. Here's an excerpt from the study:

> Survey results involving hundreds of thousands of *CR* members show that EV and plug-in hybrid drivers pay half as much to repair and maintain their vehicles. Consumers who purchase an electric car can expect to save an average of $4,600 in repair-and-maintenance costs over the life of the vehicle, compared with a gasoline-powered car.

One of the study's architects went on to state that "[I]t has long been well-known that EVs are cheaper to maintain than their gasoline-powered counterparts . . . but this is the first time we've had enough hard data from actual EV owners to prove the point." Check it out here: www.consumerreports.org/car-repair-maintenance/pay-less-for-vehicle-maintenance-with-an-ev.

Let me summarize.

The data says that you'll pay less to maintain an EV. How much less? According to the US Department of Energy:

> **Yearly average maintenance for an ICE vehicle: .101 per mile.** That's $1,000 per year for driving 10,000 miles, and $2,000 per year for 20,000 miles.
>
> **Yearly average maintenance for an ICE vehicle: .06 per mile.** That's $600 per year for driving 10,000 miles, and $1,200 per year for 20,000 miles.

You can see for yourself at www.energy.gov/eere/vehicles/articles/fotw-1190-june-14-2021-battery-electric-vehicles-have-lower-scheduled.

And again I want to emphasize that the numbers in this brief section are all exclusive of the cost of fuel.

Speaking of costs, let's now turn to the proverbial elephant in the room: forking over a mountain of cash in order to start recouping all these fuel-and-maintenance savings. And no, I'm not missing the irony at work here.

Whoa — EVs Seem Expensive! Maybe We Should Look at the Numbers

It's not exactly groundbreaking journalism to report that EVs can be expensive. But as I've outlined in the preceding section, *cars* are expensive. Even free ones are expensive to operate.

I didn't start with the bottom line sticker prices because I feel it's important to establish that point upfront. Buying *any* car is a major expense. So let's kick off this section by looking as some possibilities for getting that sticker price as low as possible.

I cover this topic in a bit more detail in the chapter about buying used, but if you're looking for inexpensive EVs, what follows here are the five least expensive models to put on your shopping list.

Note that prices are as of early 2022, and they do include the $7,500 tax credit, which absolutely should be considered when weighing the car's out-the-door price.

- **Nissan Leaf:** $20,925
- **Mini Cooper SE:** $23,250
- **Mazda MX-30:** $27,195
- **Hyundai Kona EV:** $27,745
- **Chevy Bolt:** $32,495
- **Volkswagen ID.4:** $34,455
- **Kia EV6:** $34,615
- **Hyundai Ioniq5:** $37,395

Wait — was that eight? Maybe I need help with numbers here. At any rate, as of the beginning of 2022, the average price of a new car in the US is

$47,100

I would add a sentence or two about how that number seems oppressively high, but such a sentence probably just makes me sound old. The self-evident reason I mention it is that each of the electric models I've listed here can be had for much less than the average new car price today.

To recap:

- **Don't get a car** remains the best financial move.
- **Get a free car** remains the second-best financial move.
- **And somewhere close behind free car, at least from a purely financial standpoint, comes getting a brand-new electric, choosing from the preceding list.** The new EV would be covered by warranty, and you should expect very low maintenance costs in the first five to seven years of ownership.

Bottom Line: Estimating (Using Numbers) the Total Cost of Ownership

So it's very cool to calculate all the savings you will likely enjoy as part of your EV ownership. It's even more cool to have someone else do the math for you. If you do a quick Internet search for *electric vehicle total cost of ownership calculator*, you will quickly find that several entities have been busy over the past few years doing exactly this.

One I particularly like is the EV savings calculator from Pacific Gas and Energy (known by the initials of PG&E). The way it works is that you simply click on the EV model you're considering and the website presents a bar chart showing the amount you can expect to pay for one car versus a comparable model over five years of ownership.

Regardless of where you live, every single person considering an electric vehicle should know about this tool at `https://ev.pge.com/vehicles`.

For example, clicking on the Bolt EV produces the chart shown in Figure 4-4, comparing it against a Volkswagen GTI and coming up with a very back-of-the-napkin result based on nationwide prices of gasoline. It turns out that the Bolt EV, based on current gas prices and 20,000 miles of driving per year, ends up costing $283 less per month than the GTI. That's a lot! Even if you don't pay California gas prices — maybe you're paying even higher Canadian gas prices, or less pricey Carolina gas prices — the monthly savings is just that: schmoney in the old schmocket.

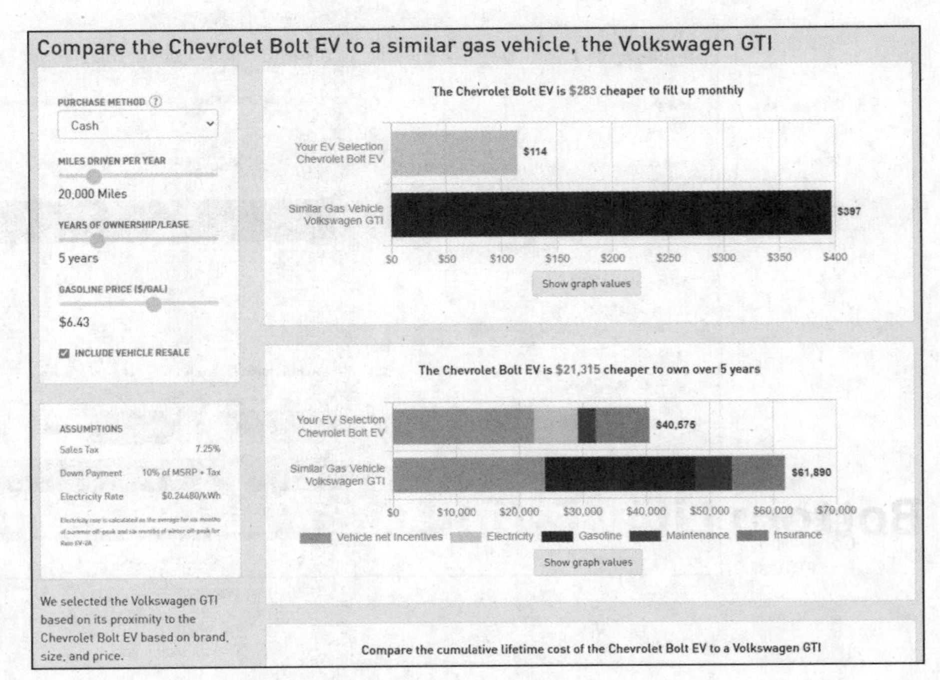

FIGURE 4-4: Comparison-shopping ICE versus EV.

If you then click the Select Electric Vehicle button, you're taken to another page, which allows you to tweak some variables based on your location. Here's where things can get interesting, because it's on the subsequent page where you find that the Bolt EV, based on current California gas prices and 20,000 miles of driving per year, ends up costing $283 less per month than the GTI. (See Figure 4-5.) That's a lot! Even if you don't pay California gas prices — maybe you're paying even higher Canadian gas prices, or less pricey Carolina gas prices — the monthly savings is just that: schmoney in the old schmocket.

When you switch from your GTI to the Bolt, you should most definitely save and invest that extra $283, or at least put that extra amount toward retiring the vehicle loan. It's also of note that $279 will get you a pretty sick pair of sneakers is all I'm saying. Maybe pay the car down at $200 per month for five months, and then, as the kids say, go ham, dude. Yolo and such. (In fairness, this is probably only what middle-aged guys *think* kids say.)

Another interesting use of this tool is to compare like to like. For example, the Hyundai Kona comes in both a gasoline variant and an all-electric model. Figure 4-5 shows what happens when you compare a Kona EV alongside a Kona ICE.

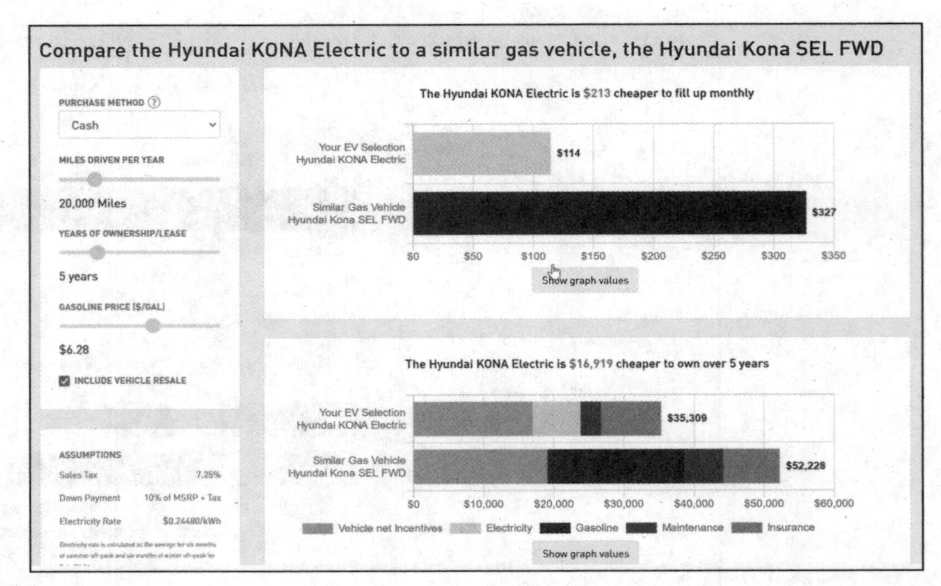

FIGURE 4-5: Comparing Konas.

Again, what I'm showing here uses California gas prices, but the winner from a cost perspective is clear: Factoring in the time value of money — which absolutely counts — the Kona EV driver will come out about $20,000 to the good versus his neighbor driving a Kona ICE.

Twenty thousand dollars is a lot of spa days for said EV driver's spouse. It's a lot of post-soccer game pizza parties for said EV driver's 12-year-old kids.

I mean, am I saying that driving an EV will make you a better spouse and parent because you won't be spending $250 per month on combustible liquid? I can't definitively answer that question. Instead, I'll have to remember to ask my wife when she returns from her weekend yoga retreat — hang on while I take a sip of tea from my World's Greatest Dad mug — and keep you posted on what she says.

» Narrowing your choices

» Evaluating used EVs

Chapter 5

Researching Your EV Options

We're still relatively early in the shift to electric transportation, but already there are some exciting selections for car shoppers, and this chapter will help you navigate your way through some of these options.

We'll begin with a look at the major categories of vehicles that use batteries for propulsion (as opposed to just powering the lights and radio; all cars and trucks have a 12V battery that handles those chores), the move towards a look at various vehicle classes — the hatchbacks, sedans, and SUVs you'll find during your exploration — and wind things up with a few tidbits that could help you take advantage of every incentive and tax credit that's available in your location.

Of course, new models of EVs are beginning to hit the roads on an almost monthly basis, so what's brand new today might be a great used car choice by the time you've got this book in your hands.

That's perfectly fine, though; I, in fact, anticipated this limitation of printed media by including a section here about buying used EVs.

Throughout, I try to keep the focus on issues that won't change — things like cargo space and performance expectations for different EV categories, or what to look for in terms vehicle (and battery) warranty.

And before we even begin, here's my parting advice for this entire chapter:

Don't overthink your EV selection.

Yes, that may sound counterintuitive from someone who's delivering an entire chapter detailing the possibilities. Yes, the selection of a car is a significant financial decision. Yes, you should try to get the most car for your money.

But ultimately, cars are inanimate objects. Yes, I have an affinity toward mine, but I also have an affinity toward cool-looking sneakers. And no matter how much I spend on the sneaker or the automobile, I try to remind myself to always ask this question: What's the worst that can happen? In the case of the ridiculously expensive sneaker, I would decide not to wear it and then I could either donate it or sell it.

In the case of the EV, it's pretty much the same thing. You might not like it for reasons that will be perfectly valid for you, or you might decide to swap out one electric vehicle for an even more powerful (or luxurious, or spacious) model. Life is short, and these are relatively minor decisions when taken into the proper context.

As long as the EV is in working condition, you can always sell it to someone else and start your shopping all over again. Trust me, with the price of fuel and the length of many of our commutes, someone will be happy to take that EV off your hands.

(Oh, and if you think you've heard the last mention of sneakers in this chapter, well then — howdy, stranger! You obviously haven't read any of my other work.)

Ok. That's probably more chapter introduction that anyone ever needs. Let's get on with the car shopping. Onward!

An Overview of the EV Market

When exploring your options, perhaps the best way to start is by identifying the different *types* of electric vehicles available for purchase. As you'll see, just because it includes the words "electric vehicle" in the description doesn't mean you won't have to keep buying gasoline.

In fact, some "electric vehicles" aren't even the kind you have to plug in so that it can charge.

All of that said, if the car has a battery, and that battery at some point supplies power to the drivetrain, then it's technically an electric vehicle.

But within that broad definition is room for lots of EV subcategories, each of which defines how often the drivetrain is powered by a battery. The following sections describe three such categories.

HEVs

HEV is the acronym for a *h*ybrid *e*lectric *v*ehicle. (The acronym made its first appearance in Chapter 1.) Any vehicle that has *hybrid* in its model name describes a car with both a gasoline motor and a (relatively small) battery that powers an electric motor.

As a rule, the power from the electric motor allows the car to use a smaller internal combustion engine, thus reducing emissions from the tailpipe.

One of the most important characteristics of a hybrid electric vehicle is that all energy for the battery comes from either regenerative braking or the gasoline engine itself. (The process is similar to how a fully ICE vehicle charges the 12-volt lead-acid battery needed for starting.)

As such, you don't power the HEV battery by plugging it in.

Hybrid electric vehicles are considered low-emission vehicles because their gasoline engines do indeed emit everything a non-EV does, although at a much lower rate.

You'll also notice that I don't/won't talk much about much hybrid electrics in this book, favoring instead coverage of fully electric vehicles. First of all, hybrids are an over 20-year-old propulsion technology, and while a hybrid car has better emissions than a traditional ICE car, if the whole world was going to go hybrid, there's been plenty of time for the transition; i.e., the "book" on hybrids was written many, many years ago.

Second, even HEVs are likely to be become functionally obsolete by decade's end — they have a combustion engine, after all, and the future of combustion engines for personal transportation isn't bright.

Do keep in mind that, as of today, HEVs remain some of the cleanest vehicles available, and because many have already been manufactured, buyers of an HEV can take comfort in the fact that their purchase isn't directly resulting in any additional carbon from product manufacturing. If you're looking to lower your carbon footprint and lower out-of-pocket expense for the vehicle, a hybrid can be a fantastic choice.

The poster child for the hybrid electric vehicle, by the way, is the Toyota Prius, a vehicle first offered to consumers in Japan in 1997, and then to the US and roughly 90 other markets in 2000. (See Figure 5-1.) As of 2017, Toyota had sold more than 6 million Prius HEVs.

FIGURE 5-1: Prius — the OG of EVs.

art_zzz / Adobe Stock

That's all well and good for Toyota, but what, practically speaking, would compel somebody to purchase a new HEV today? Well, you might consider an HEV if you determine that there's just no way you'll have regular access to an outlet for charging the battery. For many drivers, other chief motivators for buying a new conventional hybrid electric are the relative variety of options, along with the (lower) purchase price when compared to battery-only EVs.

Options

Let's take the option variety first. Compared to PHEVs and, certainly BEVs, buyers have more options for choosing a vehicle type. For example, the only way to drive an electrified pickup truck at the time of this writing is to opt for something like the Toyota Tundra Hybrid or the Ford F-150 Hybrid.

Alternatively, if you've become brand loyal to certain nameplates like Lexus and have your heart set on its RX model (the luxury, full-size SUV), the hybrid version is the only electric powertrain available.

And even though some manufacturers, like Ford and GM, have discontinued their HEV offerings, you still have some compelling *new* HEV choices from automakers like Hyundai (Elantra and Sonata), Honda (Accord and CR-V) and Toyota (Highlander, Camry, and, of course, the venerable Prius).

Cost

Now let's tackle the question of cost. As a general rule, a new hybrid electric costs less than its PHEV equivalent. For example, I'm pricing a Toyota RAV4 as a reference point and seeing that the 2022 RAV4 Hybrid (an HEV) starts at $29,075 as I write this sentence — and that the RAV4 Prime (a PHEV) starts at $39,800.

What's more, the fact that hybrids have been on the road for about 20 years now means that the used-vehicle market is similarly awash with choices. If you have a teen in your life looking for their first car to get to work and school, you can find very affordable options to get them from point A to point B, and to do so with less environmental impact than a gasoline-only car. Were I so inclined, I could set my budget with a maximum of US$10,000 and go shopping for a used 2013 Toyota Prius, a 2012 Honda Civic Hybrid, a 2016 Ford Fusion Hybrid, a 2016 Hyundai Sonata, or many others.

For virtually any car shopping budget, a hybrid option fits the bill.

TIP

Kelly Blue Book maintains frequently updated lists of hybrid cars that cost less than *x* amount of dollars. For example, typing the phrase *hybrid cars under 10K* into an online search brings this result to the top: `www.kbb.com/best-cars/best-used-hybrid-cars-under-10000`. Try it out using your own *x* amount of dollars that are within your budget.

I believe I might have mentioned this elsewhere: Batteries are expensive.

PHEVs

A PHEV is the acronym, also first introduced in Ch 1, for a *p*lug-in *h*ybrid *e*lectric *v*ehicle.

As with hybrid electric vehicles, the *hybrid* here refers to a vehicle with both an electric motor and a gasoline engine. The difference is that, with a plug-in hybrid electric vehicle, driver's will — you guessed it — plug in the car in order to charge the battery. Note that plug-in hybrids also can recharge their batteries using regenerative braking.

Whereas the typical range of a hybrid electric vehicle is just 1 to 2 miles before the gasoline engine kicks in, a plug-in hybrid electric vehicle can typically go anywhere between 10 to 40 miles before needing help from the ICE. As Chapter 4 explains in some detail, this is more than enough range for most daily driving in the United States.

You can easily handle 10 to 40 miles of range with an overnight charge in most any Level 1 wall outlet, but plug-in hybrid vehicles can take advantage of Level 2 charging as well. And if the charging level stuff is a new concept to you, please refer to Chapter 7, which is all about EV charging.

You have lots of plug-in hybrid vehicles to choose from in today's market — over 30 available in the US alone — so it's difficult to spot the pioneer of the PHEV space in quite the same way you can for an HEV.

If I had to say, though, I'd point to the Chevrolet Volt, which was brought to market in late 2010 (the first Volts were of the 2011 model-year vintage). (See Figure 5-2, whose caption is a nerdy DC Comics reference, which is in all too short a supply these days, amirite?) When the Volt was introduced, it was a standout in the hybrid space by virtue of its plug. It could go about 35 miles on its electric power alone, a range that was improved to just over 50 miles in more recent model years. The Volt won several awards, including the Detroit Auto Show 2011 North American Car of the Year, and the World Car Awards 2011 World Green accolade. In 2019, however, GM ended production of the Volt, opting to invest more time and resources into the production of the battery-only Bolt.

FIGURE 5-2: If the Batmobile is for Batman, this one is for Harvey Dent.

A. Aleksandravicius / Adobe Stock

The Chevy Volt is known as the Vauxhall Ampera in the UK and the Opel Ampera throughout the remainder of Europe.

Other prominent examples of plug-in hybrid electric vehicles are the Ford C-Max, the Fusion Energi, the Hyundai IONIQ PHEV, and the Kia Optima PHEV. All of these have been discontinued, though, so they might present an affordable electric option if you're in the used-car market. (I tell you more about that topic later in this chapter.) Once again, it's not just me who thinks that we've reached the end of an era when it comes to ICE. The auto manufacturers seem to be telling us something.

Some other prominent plug-in hybrid electric vehicles that are still in production at the time of this writing are the BMW 330e, the Audi A3 E-Tron, the Jeep Wrangler 4xe, and even the Ferrari SF90 Straddle. Exact numbers are difficult to obtain, but I'd guess that the highest-volume PHEV out there right now is the Toyota RAV4 Hybrid.

Especially if you're in Europe, my counsel is to think twice about a PHEV purchase. The EU is leading the way in terms of phasing out any personal vehicle that carries a tank of gas. For you Europeans out there, this take on the uncertain future of PHEVs might be worth a read: `https://europe.autonews.com/automakers/once-green-plug-hybrid-cars-suddenly-look-dinosaurs-europe`.

Despite Europe's dissing of the PHEV, I do understand the car's allure, especially on the wide open roads of the American Midwest and Mountain West. For now, at least, it's a have-your-cake-and-eat-it-too solution that offers the ability to go just about anywhere, especially on long road trips, without the worry of finding a fast charger and the potential delay of lengthy charging stops. (If you're on a road trip and the only option is a Level 2 charger, it can take a very long time, indeed, for a pure battery-electric to charge.)

In other words, PHEVs allow you to be ecofriendly while driving around and taking the kids to school, while still carting those same kids to hockey tournaments in far-flung locations without allowing for an hour or more of charging time to those hockey tournament road trips. Figure 5-3 shows one potential soccer-team-chauffering machine.

It is worth bearing in mind, however, that PHEVs (and HEVs) require all the maintenance of an ICE, because they are ICE cars. Gas stations, oil changes, spark plugs, timing belts . . . it's all part of the package when using a hybrid. We talk more about this very issue in Chapter 9.

FIGURE 5-3: Yes, even minivans can get into the PHEV act.

Steve Lagreca / Shutterstock.com

BEVs

The *b*attery *e*lectric *v*ehicle, of course, is the main focus of this book, and receives the most attention in this chapter.

What sets a battery electric vehicle apart is that it contains nothing that produces tailpipe emissions — it uses rechargeable batteries that draw their power either from regenerative braking (while driving) or from the electric grid (while parked).

In sum, all BEVs are *zero-emission vehicles*, and will therefore play a vital role in humanity's collective effort to keep the earth a habitable place (or at least habitable by humans). As an added benefit, BEVs tend to be easier to maintain and simply more fun to drive. If you have children right now under the age of ten, chances are rather good that those children will never drive anything other than an electric vehicle.

Looking to pinpoint the quintessential BEV? Well, as much as I might like to report, dear reader, that the leader in the EV space is General Motors, which parlayed its groundbreaking work with the EV1 into a decade of innovation and market dominance, facts simply don't support this assertion. The leader in BEVs right now is Tesla — full stop. Its 2012 Model S was the EV that put EVs on the map. (See Figure 5-4.) As of now, in early 2022, the general expert consensus is that everyone else is playing catch-up.

FIGURE 5-4: This was for EVs what the iPhone was for phones. The Tesla Model S, circa 2012.

evannovostro / Adobe Stock

Why might you consider a BEV? Let me count the ways. You'd consider an EV if you:

- Never want to stop at a gas station again
- Want a quieter ride
- Want lower total cost of ownership (TCO) over the life of the vehicle
- Want your next ICE or hybrid car to avoid becoming obso . . . — have you been paying attention? Doesn't this *entire* book address this question in some way?

REMEMBER

What I'm saying is that if you have the financing or cash that can accommodate an EV in your budget, I can't think of a reason *not* to drive a fully electric car (or truck, assuming that the truck you want/need is available).

Now that you have a general sense of what's available when you're interested in a car that's powered, in full or in part, by a battery, it's time to talk more specifically about some of the options out there.

The next section tackles that job for you.

CAN I JUST GET MY FLYING CAR ALREADY?

Ever since humankind first took to the skies, men and women have dreamed about the advent of the flying car.

My first reaction to that statement is that we have flying cars already. They're called airplanes.

My second reaction to that statement is that a flying car needs everything that an airplane needs. Until someone invents UFOs or something, we humans are bound by earthly physics, which dictate that in order for something to fly, it needs an appropriate amount of lift. In the case of a passenger vehicle (which is kind of a tube with wheels), the lift would come from either a fixed wing, which would kinda make it an airplane, or a spinning rotor, which would kinda make it a helicopter.

You probably know where I'm heading with this. Indeed, a few companies in the world are making flying car prototypes. One such company is in Slovakia, and its prototype has recently been granted an airworthiness certificate, which I guess is better than the alternative, which in turn takes its flying car concept one step closer to actually being sold to the public, or at least to very wealthy members of the public who also happen to have a pilot's license. In any event, the next time you find yourself trapped in a boring meeting, I invite you to read more about Anton Zajac and his firm's plan to bring the $700,000 AirCar to market, by taking a look here: `www.euronews.com/next/2022/01/27/this-flying-car-could-be-the-future-of-travel-after-regulators-officially-cleared-it-for-f`.

- In today's installment of "Tell me you're rich without telling me you're rich," if you possess a pilot's license yet are not paid a salary to fly aircraft, you probably have a bank account balance with several numbers to the left of the decimal point.

In any event, even if the technology is feasible, which it certainly is (the tech is in airplanes and has been for over 100 years, after all), it takes no more than about three seconds of thought before *massive* obstacles standing in the way of flying-car adoption present themselves. For example, where do you take off and land a flying car, and do you need clearance from a governing body, like the FAA? You can't just drive a car with wings down a crowded interstate highway — those other vehicles probably won't appreciate seeing 8- to 12-foot wings suddenly *sproinging* out from your door — and you probably can't reach takeoff speeds on a relatively uncrowded arterial city road.

As a purely practical matter (never mind the technology), I think cars are *much* more likely to drive themselves between destinations than they are to fly.

Separating the Wheat from the Chaff

I spell out, earlier in this chapter, the fact that electric vehicles encompass three different types: HEV, PHEV, and BEV. Time now to examine the various categories of electric cars within these spaces. If that previous sentence sounds too convoluted for its own good, what I'm referring to here are the different vehicle body styles.

Are you in the mood for a hatchback? A sedan? An SUV? Something else entirely? Whatever it is, there's likely an option for you that either won't produce much in the way of CO2 emissions or won't produce anything at all.

Hang on a sec! Just slow your roll there, bub!

Before you skip right to the specs on the car of your dreams, just know that this is a *significant* list, but it's also a *curated* list. I'm presenting you with the most compelling options in each of the major automotive classes, and the bias of the list curator says that BEVs are preferred over PHEVs, which in turn are favored over HEVs. More to the point: HEVs don't even make the list, and only a select few PHEVs garner attention in this space.

BEVs are preferred in this chapter, and, hopefully, for your situation because a) they have no tailpipe emissions and b) as I discuss earlier in this chapter with regard to ownership cost, a price difference of $5,000 to 10,000 between a BEV and an HEV is more or less a wash over five or six years of ownership.

Also, I don't include manufacturer websites here, because I trust your online search skills as much as I trust my own. And I'll update this list whenever the book is revised, which, if you order a box or two as gifts for friends and family, will be frequently.

I've broken the sedan category into two sections: midsize sedan and luxury/performance sedan. Sure, the breakpoint between the two isn't cut-and-dried, but it's my list, and I say that roughly US$55,000 marks the dividing line between performance/luxury and midsize. I know what you're likely thinking. I just work here; I don't control the prices.

Now then: Let's go car shopping!

All prices shown here are exclusive of the US federal tax credits because they certainly won't apply anywhere except in the US. Most BEVs still qualify for $7,500 in federal tax credits, with two main exceptions: Chevy Bolt and anything from Tesla.

Hatchbacks

The French came up with the first hatchback back in 1938 — Citroen's 11CV Commerciale. Much has changed since then.

Chevrolet Bolt EV

Chevrolet Bolt is an impressive hatchback, so much so that it earned the 2017 Car of the Year award from *Motor Trend*. (See Figure 5-5.)

In 2022, the Bolt received an exterior and interior refresh, while lowering the manufacturer's suggested retail price (or MSRP). Unfortunately, the power train is a bit lacking when compared to other EVs in the small car/crossover hatchback segment, and the car has been plagued in recent months by a battery recall due to the risk of thermal events. (For those skipping about like this book was the Internet, thermal events are covered in Chapter 3.)

In addition, the Bolt can only charge at 55 kW, compared to a Model 3's 250 kW fast charging capabilities. This means that a Bolt probably isn't a great fit for the weeklong road trip. (If there's a Level 2 charger at your destination and you can top up overnight on the other hand, you are probably in great shape.)

The great news on the Bolt EV front is that it's one of the most affordable ways to buy a new EV. While I was doing final revisions, Chevy announced a price cut on its Bolt EV and Bolt EUV hatchbacks. This means that you might be able to grab a capable EV for much less than the ones on offer from competitors.

Bolt EV range: 259 miles

Price: starting at $25,600 for the EV, and $27,200 for the larger EUV

Hyundai Ioniq 5

Hyundai's newest and most eye-catching of its Ioniq lineup, the 5 comes in two varieties: a single-motor rear-wheel drive and a dual-motor all-wheel drive. (See Figure 5-6.) It offers two battery pack options as well — one with 58 kilowatt-hours (kWh) of juice and the other storing 77 kilowatt-hours, with a top EPA range of 303 miles.

EPA combined range: 303 miles*

Base price: $44,875*

FIGURE 5-5: The Chevy Bolt EV.

dennizn / Adobe Stock

FIGURE 5-6: The Hyundai Ioniq 5.

Mike Mareen / Adobe Stock

Mini Cooper SE

For my money, the Mini Cooper looks like it has always been an electric vehicle; I'm surprised these things *ever* had a gasoline engine.

The Mini Cooper is a stylish, versatile city driver (se Figure 5-7), but with a range of just 114 miles, it's not quite cut out for summer road trips to the Grand Canyon and back. If you live anywhere throughout most of Europe or the closely connected cities of the Northeast US, this might be a perfect car for you. If you live in suburban Tulsa, you might want to consider other options, or at least consider the Cooper as a second car.

Signature: 114 miles, $30,750

Signature Plus: 114 miles, $34,750

Iconic: 114 miles, $37,750

FIGURE 5-7: The Mini Cooper SE.

artographer34 / Adobe Stock

Nissan Leaf

The original Nissan Leaf, introduced in 2010, featured a range of just over 70 miles. Unfortunately, some of this perception of limited range has been hard to shake, as the Leaf probably doesn't get the attention it deserves for being a groundbreaking car that has continued to improve over time.

If you want an economical, everyday commuter that can still haul most IKEA furniture, newer Leaf hatchbacks are more than up to the task. (See Figure 5-8.)

Nissan Leaf SV: 149 miles, $35,115

Nissan Leaf S Plus: 226 miles, $39,125

Nissan Leaf SL Plus: 215 miles, $45,520

FIGURE 5-8: The Nissan Leaf.

VanderWolf Images / Adobe Stock

Volkswagen ID.3

The ID.3 was first delivered to German customers in September 2020, and it quickly ranked among the world's top ten best-selling plug-in cars for both 2020 and 2021. (See Figure 5-9.)

The Pure Performance style is intended to be Volkswagen's budget electric vehicle. The Pro S model offers a bigger battery and more range, but costs about $15,000 more. The ID.3 won't be sold to the US market.

Volkswagen ID.3 Pure Performance:

Range: 210 miles, 55 kWh battery

Base price: $39,000

Volkswagen ID.3 Pro S:

Range: 340 miles, 82 kWh battery

Price: $54,000

FIGURE 5-9: The Volkswagen ID.3.

Frank Gärtner / Adobe Stock

Midsize sedans

This one's for the family with 2.5 kids. (That's the average number of kids in a US household, which is actually an old stat. The current number is 1.9.)

Polestar 2

If you're wondering why it's the Polestar 2, you're not alone. The Volvo-owned Polestar 1 one was a Super GT plug-in hybrid (GT usually refers to Grand Touring when part of a car model name), but that car has since been discontinued.

The Polestar 2 is a battery-only car, and it has two variants: a single-motor front-wheel drive and a dual-motor all-wheel-drive. It looks very sporty, performs well, and can hold lots of your stuff. (See Figure 5-10.)

Polestar 2 Long-Range Single Motor: 265 miles range

Price: $47,200

Polestar 2 Long-Range Dual Motor: 249 miles range

Price: $51,200

FIGURE 5-10: The Polestar 2.

VanderWolf Images / Adobe Stock

Tesla Model 3

The Model 3 is the planet's best-selling battery-electric vehicle, and for many reasons. For starters, the car keeps improving, even though these improvements aren't immediately apparent at a glance — the 2022 model looks almost exactly like the 2017 version. (See Figure 5-11.)

Most of the improvements have been in the form of more energy-dense battery packs, along with efficiency improvements, like the use of a heat pump. When you factor in resale value, the Model 3 is probably the most economical vehicle you can now purchase.

Model 3 (RWD) range: 272 miles

Price: $46,990

Model 3 Long Range range: 353 miles

Price: $54,490

Model 3 Performance range: 315 miles

Price: $61,990

FIGURE 5-11: The Tesla Model 3.

Mike Mareen / Adobe Stock

Luxury/performance sedans

In the electric vehicle space, electric sedans and performance sedans or almost always one of the same. The big batteries mean a lot of raw power, a lot of range and a lot of make-sure-your-seatbelts-are-fastened launches away from stoplights.

The reason why you're seeing a lot of luxury sedans right now has to do with selling an electric vehicle at a profit. It's not the way I wish things were, and I wish governments the world over (and especially in the US) would encourage EV adoption the way they've encouraged fossil fuel exploration or highway construction, but the lay of the land is such that EVs are expensive to manufacture, so automakers are more-or-less forced to find buyers who can afford premium items such as luxury sedans.

Plus, you get leather (or vegan leather) seats.

Audi E-Tron

You may remember the E-Tron making an appearance as Tony Stark's car of choice in 2019's *Avengers: Endgame*, although for some reason the sound mixers decided that an electric car needed the sound of a gas engine, or else audiences would . . . get confused? I don't know.

What I do know is that Audi and Porsche share the same platform and power train, and both nameplates offer sleek-and-stylish electric sedans at roughly the same price points. (See Figure 5-12.)

Audi E-Tron GT range: 238 miles

Base price: $103,445

Audi RS E-Tron GT range: 232 miles

Base price: $143,445

FIGURE 5-12: The Audi E-Tron.

VanderWolf Images / Adobe Stock

BMW i4

The BMW i4 is the company's first fully electric sedan. The initial production run comes in two flavors: a single-motor eDrive40 and a dual-motor M50. (See Figure 5-13.)

BMW enthusiasts will likely recognize the M as an indicator of BMW's sports-performance subsidiary:

i4 eDrive40 with 18-in wheels: 301 miles, $56,395

i4 M50 with 19-in wheels: 270 miles, $66,895

FIGURE 5-13: The BMW i4.

Santi Rodríguez / Adobe Stock

Lucid Air

I just happen to know someone who's familiar with the Lucid Air, and the same person, if asked, will tell you about the sedan's incredible aerodynamics and efficiency, both of which help make the Air the reigning champion in terms of BEV range. (Okay, that someone is me, but I trust the guy; he won't steer you wrong — just see Figure 5-14.)

Speaking of range, the Air Dream edition, at $169,000, is more than twice the price of the Air Pure, which can be had for a "mere" $77,000. The Pure still offers more than 400 miles of range.

Air Pure range: 406 miles

Price: $77,400

Air Touring range: 406 miles

Price: $95,000

Air Grand Touring range: 516 miles

Starting Price: $139,000

FIGURE 5-14: The Lucid Air.

Mike Mareen / Adobe Stock

Mercedes-EQ EQS

From the nameplate most closely associated with luxury sedans comes the EQS, a luxury *electric* sedan. (See Figure 5-15.)

The EQS interior is peppered wall-to-wall with screens and other luxury technology, as you can see in the figure. You can't be blamed if, while sitting in an EQS, you think you're piloting a spaceship:

EQS450+ range: 350 miles

Price: $103,360

EQS580 4Matic range: 340 miles

Price: $120,160

Porsche Taycan / Taycan Sport Turismo / Taycan Cross Turismo

I haven't yet ridden in one but, by all accounts, the Porsche Taycan is one of the best-driving EV out there today. (See Figure 5-16.) As you can see from the following list, you also have about 38 varieties to choose from (or something like that).

FIGURE 5-15: The Mercedes-EQ EQS.

Julia Lavrinenko / Adobe Stock

Besides the trademark Porsche performance, the Taycan also boasts of DC fast-charging times that are industry leaders (although Lucid has recently bested Porsche in this category); packs can reportedly charge from 5 percent to 80 percent in just 23 minutes.

The downside of owning a Porsche is that the charging might be needed more often, as Taycan ranges certainly leave something to be desired:

Taycan (RWD) Performance Battery: 200 miles, $84,050

Taycan 4S Performance Battery Plus: 227 miles, $110,720

Taycan GTS: 215 miles, $132,750

Taycan Turbo: 212 miles, $152,250

Taycan 4S Cross Turismo: 215 miles, $111,650

Taycan Turbo Cross Turismo: 204 miles, $153,850

Taycan Turbo S Cross Turismo: 202 miles $188,950

Tesla Model S

I name-check the Tesla Model S throughout this book, so you probably already have a good sense of its awesomeness. (See Figure 5-17.) It's really fast, it's expensive, it no longer has a steering wheel, and — oh, by the way — it's insanely fast:

Model S Long Range range: 405 miles

Price: $96,190

Model S Plaid with 21-inch wheels range: 348 miles

Price: $135,690

FIGURE 5-16: The Porsche Taycan.

VanderWolf Images / Adobe Stock

FIGURE 5-17: The Tesla Model S.

Mike Mareen / Adobe Stock

Crossovers

A crossover — sometimes called a compact SUV — is for folks who like an SUV but hate the fact that most SUVs are built on top of a pickup-truck frame. A crossover aims to please by ditching the truck frame and going with a passenger-car frame.

The net is that you typically get a nimbler ride with a crossover, better efficiency when compared to full sized SUVs, and still have plenty of room for all of your kids, dogs, groceries, and luggage. They've become *extraordinarily* popular body style, especially in the US, especially given the relatively short time they've been on the market.

Kia Niro EV

This front-wheel-drive car offers an attractive mix of space, range, and affordability. (See Figure 5-18.) And it's still pretty quick for a single-motor EV, capable of a 0-to-60 time of 6.5 seconds:

EPA combined range: 239 miles

Base price: $41,205

FIGURE 5-18: The Kia Niro EV.

Santi Rodríguez / Adobe Stock

Tesla Model Y

Tesla's wildly popular crossover SUV is, well, a cross between the Model 3 and the Model X SUV. If you want a conventional compact SUV for comparison, it's a little bigger (5 inches longer and 3 inches wider) than a Honda CR-V. (See Figure 5-19.)

Both current models feature all-wheel drive, thanks to dual motors, one at each axle:

Model Y Long Range range: 318 miles

Model Y Performance range: 303 miles

Base price: $62,990

FIGURE 5-19: The Tesla Model Y.

CenturionStudio.it / Adobe Stock

Volkswagen ID.4

The ID.4 is Volkswagen's first EV sold in the United States and is also the first fully electric crossover offered by VW. (See Figure 5-20.)

The World Car Awards named it the World Car of the Year in 2021:

EPA combined range: 260 miles

Base price: $41,955

FIGURE 5-20: The Volkswagen ID.4.

VanderWolf Images / Adobe Stock

Ford Mustang Mach-E

Ford began delivering its first fully electric vehicle in 2021, and the power it delivers certainly lives up to the Mustang branding. (See Figure 5-21.)

In fact, the Mach-E GT Performance produces more torque than the (much more expensive) V-8 Shelby Mustang GT500. I don't know what V-8 refers to, but my guess is that it means loud and slow.

EPA combined range: 305 miles

Base price: $44,995

Kia EV6

The Kia EV6 is the newer, more stylized version of the Kia Niro, although it's priced at only $100 more than the Niro, as of this writing. (See Figure 5-22.)

FIGURE 5-21: The Ford Mustang Mach E.

Belogorodov / Adobe Stock

The EV6 includes a version with a 58 kWh battery pack and a pricier, rear-wheel drive version with a 77 kWh pack capable of 310 miles of range.

EPA combined range: 232 miles

Base price: $42,115

FIGURE 5-22: The Kia EV6.

Mike Mareen / Adobe Stock

Toyota RAV4 Prime

The RAV4 is the most popular crossover in the US market, and the RAV4 Prime is an attractive option for those who are wanting a) the Toyota badge, b) the popular RAV4 model, and c) an electric range of 42 miles (plenty for most daily commuting). (See Figure 5-23.)

As a bonus, the RAV4 qualifies for the full $7,500 tax credit because of its battery size (I tell you more about that topic later in this chapter):

EV-only range: 42 miles

Base price: $41,015

FIGURE 5-23: The Toyota RAV4 Prime.

vivoo / Adobe Stock

SUVs

These are for the folks who don't mind the pickup truck frame.

Volvo XC60 Recharge PHEV

This model is Polestar's luxury smaller SUV (when compared to the XC90). (See Figure 5-24.) It has a very limited EV range of 19 miles, but the car remains a sporty option for someone looking for something that can cart the kids to school and back and then haul them to a soccer tournament in the neighboring city, all in the span of eight hours:

EV-only range: 19 miles

Base price: $55,345

FIGURE 5-24: The Volvo XC60 Recharge.

franz12 / Adobe Stock

Jaguar I-PACE

The Jaguar I-PACE debuted back in June of 2018, and in 2019 was named the World Car of the Year at the 2019 World Car Awards, conducted each year that the New York Auto Show. It was the first Jag to win this award.

The I-PACE received a refresh for model year 2022, and now comes standard with a 90 kWh battery pack and two electric motors capable of hitting from 0-to-60 miles per hour in less than five seconds. (See Figure 5-25.)

EPA combined range: 234 miles

Base price: $71,050

FIGURE 5-25: The Jaguar I-Pace.

Dmitry Dven / Adobe Stock

Tesla Model X

The Model X was Tesla's second mass-market car, and its first mass-market SUV. It's probably most famous for its DeLorean-esque falcon-wing doors, not to mention its 0-to-60 mph times of fewer than three seconds in the fastest models. (See Figure 5-26.)

They aren't cheap, but then again, they accelerate faster than most Ferraris, and Ferraris can't haul around mountain bikes on a tow hitch:

Model X range: 332 miles

Price: $114,990

Model X Plaid with 22-inch wheels range: 311 miles

Price: $138,990

Rivian R1S

The Rivian R1S boasts an available three rows of seating, an adventure package (premium interior, upgraded sound system, and so on) and the striking headlights shared by its pickup truck sibling. (See Figure 5-27.)

FIGURE 5-26: The Tesla Model X.

franz12 / Adobe Stock

With a large 135 kWh battery pack, the R1S is capable of 316 miles on a full charge.

EPA combined range: 316 miles

Base price: $76,575

FIGURE 5-27: The Rivian R1S.

Mike Mareen / Adobe Stock

Porsche Cayenne E-Hybrid

Porsche released a new, improved PHEV version of its popular Cayenne line of SUVs in 2021 with a lager battery that in turn gives these models a higher battery-only range, although extracting battery-only range numbers from the Porsche website is a nightmare.

As with almost any Porsche, you can expect a mix of driving performance and luxury fit and finish that have become synonymous with the Porsche badge (it's a Volkswagen luxury brand; see Figure 5-28).

EV-only range: 17 miles (Cayenne), 15 miles (Cayenne Turbo S)

Base price: $84,650 (Cayenne), $166,650 (Cayenne Turbo S)

FIGURE 5-28: The Porsche Cayenne E-Hybrid.

franz12 / Adobe Stock

Pickups

If you're shopping for a pickup today, chances are good that your only option is a hybrid pickup.

Rivian R1T

Rivian lays claim to the First Mover title in the US electric pickup truck market with its R1T truck. It has also been awarded Motor Trend's 2021 Truck of the Year and has also earned a Car and Driver Editor's Choice award.

As a mid-sized pickup (Rivian is marketing it as a "lifestyle" truck geared towards the camping set, or at least the camping set amongst Silicon valley tech workers), the R1T is bigger than a Ford Ranger, but smaller than something like a Ram 1500 or Toyota Tundra. (See Figure 5-29.) Acceleration is comparable to performance electric sedans, achieving 0-60 speeds of around 3 seconds, so make sure to bungee down all that camping gear.

Like the R1S, the R1T comes with a rather massive 135-kWh standard battery pack, with plans for a 180-kWh pack in 2023. Rivian claims that the R1T can tow more than 11,000 pounds.

It's also exceedingly difficult to figure out what the R1T's price is unless you go in and configure one. The starting price model of $67,500 doesn't exist at the time of this writing. By the time you've selected what are fairly standard options, it looks like you'll be much closer to spending $90-$100k.

EPA range: 314 miles

Price: $67,500, but as much as $120,000 when equipped with all options.

FIGURE 5-29: The Rivian R1T.

Tada Images / Adobe Stock

Ford F-150 Lightning

The Ford F-150 Lightning takes the best-selling F-150 ICE truck body and switches out the gasoline engine for a pair of electric motors and a battery pack that comes in one of two sizes: 98 kWh or 131 kWh.

The result, according to those who've had a chance to drive what should prove to be one of Ford's most breakthrough vehicles this side of the Model T, is a zero-emission pickup that's both quick and capable — able to tow up to 10,000 pounds. (See Figure 5-30.)

In addition, the F-150 can supply power to the house. In the event of a power outage, this would allow a fully charged truck to supply up to three days of electricity for the average US household. (I discuss vehicle-to-home charging opportunities and challenges in Chapter 10.)

> **EPA range:** 230 miles (Standard Range pack) or 320 miles (Extended Range pack)
>
> **Price:** starts at $41,769, but up $92,669 to when selecting the Platinum Edition with Extended Range battery

FIGURE 5-30: The Ford F-150 Lightning.

Mike Mareen / Adobe Stock

Tesla Cybertruck

If you're reading a book about electric vehicles and you don't know what the Tesla Cybertruck looks like already, I'm here to call into question your EV book-reading bona fides. Whatever else you think of Tesla or Elon, the Cybertruck has been wildly successful at generating attention. (See Figure 5-31.)

The Cybertruck won't be produced until after this book goes to print, which is probably appropriate because the thing *looks* like it comes from the future. (Full disclosure: Our family has a reservation.)

At the time of the reveal event, the company's owner, Elon Musk, mentioned a starting price of $39,000 for the most entry-level model, with the high-end, tri-motor model clocking in at $69,000. I'm highly doubtful that either of these price points will hold by the time they roll off the lines in (for now) 2023. At the earliest.

It is purported to tow up to 14,000 pounds (putting it in league with large pickups like the Ram 1500) and will have an estimated range of 500-plus miles using the largest battery pack. (As with the announced price, time will tell if that 500-mile range claim holds true.) Tesla says the top-end variant will launch from 0-60 at 2.9 seconds.

EPA range (claimed): 250 miles (Single motor) or 500 miles (Tri motor)

Price (announced): $39,900 for the single motor, 49,900 for the dual motor, and $69,900 for the tri motor

FIGURE 5-31: The Tesla Cybertruck.

Mike Mareen / Adobe Stock

GMC Hummer

Even though the Hummer isn't technically a pickup truck, it's certainly not a car, either. (See Figure 5-32.) In fact, the EPA has labeled it a heavy-duty vehicle. Because it's not a very good pickup truck, and because it's not a very good electric vehicle, I'm of the opinion that this vehicle shouldn't even exist.

Yet it does, and yet there's a line out the door of customers wanting to get their hands on one, something that's not helped very much by GM's ability to supply these things. In all of Q1 2022, GM delivered only 99 Hummer EV units.

Because of such limited supply (and because of a lot of other factors that I, due to time constraints, couldn't unpack were I given the entire book to do so), someone who should be putting their money to better use (says the guy who spends way too much on sneakers) apparently paid $255,000 at a Texas dealer auction. (Don't believe me? Check out `www.reddit.com/r/electricvehicles/`.)

If you *really* like Hummers, know that the only new Hummer you can buy today is this 47 MPGe beast. (MPGe was a topic discussed in Chapter 4, and MPGe is quite low as far as these things go. A Model 3 is rated at over 130 MPGe.)

EPA combined range: 329 miles

Base price: $108,700

FIGURE 5-32: The GMC Hummer.

Mike Mareen / Adobe Stock

Others

Thank goodness there's always an Others category, perfect for housing those oddball items that don't fit anywhere else.

These Other vehicles can also be great for families with small children (the minivan) or for families seeking out some off road adventures (the Jeep).

Chrysler Pacifica Hybrid

Until the Volkswagen ID. Buzz comes along, this is the only option available for those seeking the space and versatility of a minivan. (See Figure 5-33.)

Besides all the other practical reasons to consider buying a minivan, the Pacifica comes with a 16 kWh battery pack that takes only about two hours to charge on a 240V outlet, which can provide about 32 miles of electric-only range:

EV-only range: 32 miles

Base price: $49,095

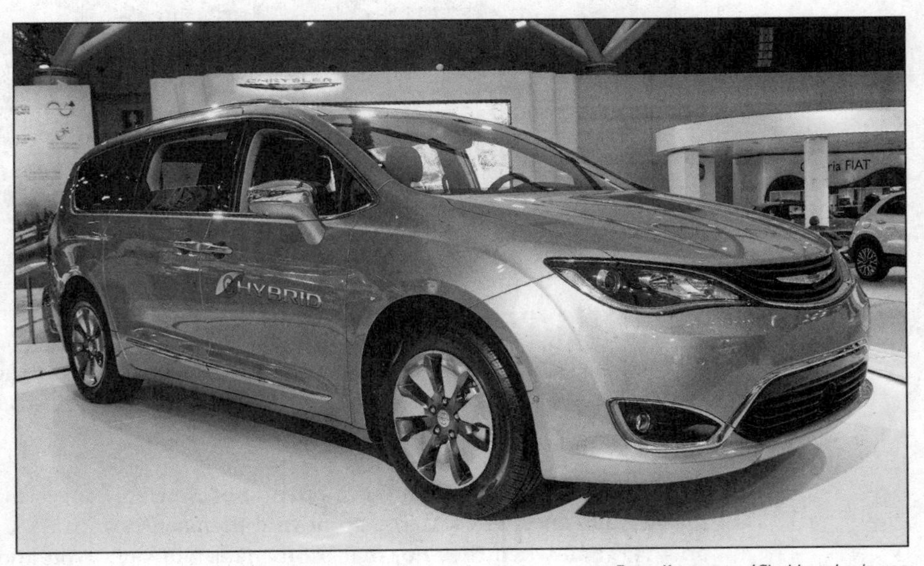

FIGURE 5-33: The Chrysler Pacifica Hybrid.

Zoran Karapancev / Shutterstock.com

Jeep Wrangler 4xe

I have a neighbor who has one of these, and it's kind of a cool sight — and sound — to watch him silently pull a Jeep into his garage.

It looks like a regular Jeep in every way (see Figure 5-34), and word is that the high-voltage pack is sealed and waterproof, making it perfectly safe for off-roading:

EV-only range: 22 miles

Base price: $54,125

FIGURE 5-34: The Jeep Wrangler 4xe.

gpriccardi / Adobe Stock

Rimac Nevera

Yes! I'm including a $2.4 million car in the rundown. This is the Other category, after all, and I won't let a car with <checks notes> over *nineteen hundred* horsepower go unmentioned. (See Figure 5-35.)

Why? Because some San Fran tech bro and/or New York investment banker in my reading audience will at some point find themselves with $2.4 million to spend on a car, and when they do, that bro/banker will say, "You know what? I really appreciate learning about the Rimac Nevera from that *For Dummies* book by Brian Culp. I mean, I was going to go out and waste that money on — I don't know — a Banksy crayon drawing or a private island in the Maldives, just to impress my 24-year-old au pair and now third wife, but instead I'm going to spend that $2.4 millie on myself. Man's gotta practice some self-care once in a while. YOLO!"

Only 150 Rimacs are planned for production, so I'll let you tech bros arm-wrestle the investment bankers over your place in line:

Nevera range: 270 miles

Price: $2.4 million

FIGURE 5-35: The Rimac Nivera.

Vova / Adobe Stock

Volkswagen ID. Buzz Microbus

The retro design of the ID. Buzz is a nod to Volkswagen's iconic VW Microbus of the 1970s, almost always portrayed with a surfboard attached to its roof or smoke rolling out of its windows — or both. (See Figure 5-36.)

If this thing ever gets produced — it's been promised ever since it appeared in a 2017 Super Bowl ad, and it's now supposed to arrive in 2024 — people will flock to it in droves.

I mean, look at how cute that thing is! Holy cow, will that thing ever come in handy for middle-aged men heading out for a weekend of camping or bringing the wife to an entirely age-inappropriate EDM festival. Should people old enough to be grandparents attend events like Ultra Miami, Tempe's Body Language, or Coachella? Well, let's get a favorite author of yours behind the wheel of an electric bus and find out! Gotta dance like there's no one watching sometimes, you know?

Range: Approximately 260 miles

Base price (estimated, and highly unlikely to happen): $40,000

FIGURE 5-36: The Volkswagen ID. Buzz.

VanderWolf Images / Adobe Stock

Deciding When (or Whether) to Buy a Used Electric Car

Although the math on the total cost of ownership (TCO) has shifted in recent years, buying a used car remains the most budget-friendly way to do your car shopping.

So, if the primary factor in your EV selection is the out-the-door price, the used market is where you'll want to spend your time. Buying used might not be the lowest TCO available today, especially if you end up replacing the battery (which may never be necessary), but car payments are car payments, and every budget has its ceiling. In most every case, buying used is the least expensive way to get behind the wheel when considering like-for-like — a used Nissan Leaf will (almost) always cost less than a new Nissan Leaf.

TIP

I know of a few sightings in recent months of used EV models (especially Teslas) being even pricier than new ones. I know: It doesn't make sense to me, but neither does having breakfast cereal for dinner, yet that too is a thing.

Apparently, folks are willing to pay to get their used Tesla *now* rather than wait six months for a new one to roll off the factory floor. Check out this article at `https://insideevs.com/features/517579/tesla-model3-used-prices`.

That said, here are a few considerations if you decide to go shopping for any of the cars mentioned in the previous section (or a different one!) by searching the classifieds. Or, if you're not my 84-year-old dad, it's more conceivable that you'll start your search from the comfort of your phone.

For starters, the biggest advantage you can bring with you in your used-car shopping is to know exactly what you're looking for. A bit of homework with an iPad or a 400-page paperback can often save you days of headache when making your purchase. The more specific you can be before the act of physical shopping begins, the more likely you are to find a car that makes you excited about signing the loan papers (or the check). Knowing that the best car for your tastes and your family's lifestyle is a RAV4 Prime, for example, saves you a lot of mental overhead, and will help you either wait for the right vehicle to come along or pounce on it when it appears.

In terms of other used-EV buying tips, take to heart the advice proffered in the next few sections.

Verifying battery life

Lithium-ion (Li-ion) batteries are miracles of the modern world — I'm jamming to the Pandora station in my wireless headphones right now, thanks to that miracle — but they *do* degrade over time nonetheless. You can do a couple of things to check on battery degradation, however.

Some cars, such as the Nissan Leaf, have readouts that tell the owner how much the battery has degraded.

What's more, every electric car stores data about battery health, and you can find several apps to help you access this data to gain valuable insight (literally, because buying a car with battery degradation issues is a five-figure mistake) into how much life remains in that pack.

Here's a listing of some of these tools:

- **Recurrent**: Recurrent is a website and app that lets you connect your EV to its servers. The company will then monitor four data points — charging status, battery level, odometer, and range estimate — in order to build a monthly health report.

  Both shoppers and dealers can also use the service to analyze cars being bought and sold — sellers can obtain a one-time battery vehicle health report.

  You can find out more about the company and the reports at their website, `https://www.recurrentauto.com/`.

* **LeafSpyPro:** This Android app plugs into a Leaf's CAN bus and provides the battery's percentage state of charge (SOC), info on the voltage of each of the battery pack's 96 cell pairs, watt-hours used by the Leaf since the last reset, and the distance to empty, based on personal efficiency settings.

TECHNICAL STUFF

What's a CAN bus? Almost all of today's cars feature electronic control units (ECUs). The wires within the car that allow these ECUs to communicate with one another is called the Controller Area Network, or CAN, or CAR-CAN.

When cars are connected to the Internet, this CAN bus can be accessed remotely, which is necessary for tasks such as remote diagnostics and over-the-air updates.

REMEMBER

The same connectivity does expose such a car to the possibility, however remote, of being hacked. Control the ECUs and you control the car.

* **ScanMyTesla:** The ScanMyTesla app is similar to Leaf Spy in that it lets users view real-time data about battery power, temperatures, lifetime stats, and capacity estimates, along with additional data about drive unit power, torque, and horsepower.
* **TM Spy:** Almost identical to ScanMyTesla in terms of function, this app provides info about individual cell voltages, lifetime battery discharge, mileage driven on the current battery pack, and more.

Digging into the exact mechanics of how to set up these apps is beyond the scope of this book, but it's worth a quick note that, with the exception of Recurrent, each of these tools (sorta) plugs into something called the ODB2 port on the car. What's that?

If you've ever taken a car in for service after seeing the Check Engine light, the technician plugs the diagnostic thingamajiggy into an ODB2 port. (More specifically, the thingamajiggy is an ODB2 scanner.) In the case of the apps just listed, most configurations use a Bluetooth adapter on the port, which in turn send the data wirelessly to the phone.

On the YouTube channel Out of Spec Reviews, host Kyle Connor talks extensively about how he used the ScanMyTesla app to track battery degradation after 100,000 miles from his Model 3, a range that required about 40 megawatt-hours of energy stored and discharged over several years of use.

It's an almost 40-minute video, but then again, I guess that's why Google has video speed controls: `www.youtube.com/watch?v=p9R8HXSnD5Y&t=923s`.

What's the bottom line? Kyle's Model 3 lost about 12 percent of its range over the course of 100,000 miles. A car with a range of 310 miles became a car with a range of 272 miles over the course of those 100,000 miles.

REMEMBER

Trust but verify. Before signing over the check, that is. When circling the car that you think is the right one, just remember that no two battery packs are exactly alike, and it pays to check on how much the battery has degraded.

One of the easiest ways to verify is simply to charge the car to 100 percent and then take it for a 50-mile test drive (keeping in mind, of course, that range is affected by factors such as temperature, wind, and speed — as with a gas-powered car, you'll go farther at 55 mph than at 85 mph).

Sure, it's back-of-the-napkin stuff, but a car that has lost more than 15 percent of its range in the first two or three years might be worth some further digging. In other words, if 300 miles when new has become fewer than 255 while still in the first few years of ownership, it might raise an eyebrow.

Start practicing Zen meditation

Or at least start practicing patience, especially when shopping used.

This advice can vary, depending on the model you have in mind, of course — a Toyota Prius is easier to find than a Toyota RAV4 Prime — but patience during the search for your EV is indeed a virtue.

In a word, it can be difficult to find the battery-powered car you deemed ideal, especially if you have other qualifying criteria, such as a specific color, model year, or mileage threshold.

The issue is simply one of quantity. The US has roughly 290 million registered vehicles cars on the road, and only 2 million or so are electric. What's more, most of those EVs are newer models, and both these factors mean that there are far fewer EVs to choose from right now.

REMEMBER

There's a tendency — at least in the US — for new-car dealers to steer shoppers away from EVs, mainly for reasons related to familiarity. (You could offer your local dealer a copy of this book to bring them up to speed, one imagines.) And, if that's the case with new-car recommendations, you can probably imagine that used-car dealers will be even less familiar with — and therefore less likely to educate you about — a used EV.

TIP

Given the paucity of EVs out there — both new and used — it pays to let your fingers do the walking through various dealers' websites *before* heading out to a car lot. It's not realistic to expect your local Ford dealer to have five used Mach-Es juiced up, awaiting a test drive.

Living with no (or very few) incentives for used vehicles

With notable exceptions (like GM and Tesla), most new electric vehicles are eligible for purchase incentives from federal and state governments. And this money counts, especially because it's usually a tax credit.

For example, a $7,500 discount on a Hyundai Ioniq5 equates to roughly $13,000 in pretax income that would otherwise have to be earned to make a $7,500 down payment. The bottom line is that the tax credit offsets roughly 15 percent of the price of an Ioniq5.

TIP

Where's the best place to buy a used electric car? In the US, it's the state that sells the most electric cars: California. Otherwise, the highest number of used EVs tends to be in the states with the highest rates of adoption, like Colorado, New York, Washington, Texas, and Florida. Yes, Florida.

If you're adjacent to one of those states, it might be worth crossing state lines to grab the EV equivalent of some OG Breds. (Sneakerheads are now nodding in deep appreciation of the *OG Bred* reference, which describes a hard-to-find pair of Nike Air Jordans that use black and red. Get it? Bred.)

However, those incentives are meant to get *new* vehicles on the road, which is as it should be, I suppose. I think it's generally a good use of government resources to incentivize electric vehicle manufacture, along with helping a buyer get an EV into circulation.

But wait! Don't despair just yet, my dear used-EV shopper. It never hurts to check for state incentives. I strongly recommend hunkering down and doing some research.

California, for example, offers programs like the Clean Vehicle Assistance Program, which offsets up to $5,000 of an eligible used EV. Colorado also offers $2,500 in addition to any federal incentives, and $5,000 for a medium0duty electric truck. Germany offers something called an *Umweltbonus*, which can apply to used electric or PHEV cars. (Like *schadenfreude*, Germans have a word for everything — *Umweltbonus* is their way of saying "environmental bonus.") The Netherlands also offers used-EV incentives, or at least did at one time.

I'm saying it's worth checking, yeah? I'm not going to research every nation, province, and state on earth.

HOW DO THOSE FEDERAL TAX CREDITS WORK, EXACTLY?

If you read Internal Revenue Code Section 30D, which you absolutely should (duh), you'll find that electric drive motor vehicles, including passenger vehicles and light trucks, are eligible for a tax credit of up to $7,500.

The amount of the credit is actually $2,500, *plus* an additional $417 for every kilowatt-hour of battery capacity in the vehicle over and above 5 kilowatt-hours. So, most plug-in hybrid electric vehicles — and certainly all battery electric vehicles — are eligible for the maximum credit allowed, which is $7,500.

Additionally, the credit begins to phase out once the manufacturer has sold 200,000 vehicles that qualify for this tax credit. (The credits are available for vehicles sold for use in the United States.) Tesla and GM have surpassed that 200,000-vehicle threshold, but it's still available to buyers of almost any other automaker, although if you're in possession of a first edition of this book, you may want to act quickly: Ford, Nissan, and Toyota are expected to cross this 200,000 limit in 2022 or 2023.

Here are two other items of note about the US federal tax credit:

- You still pay full retail price for the vehicle and then claim the tax credit when you file your taxes.
- The credit still applies when you lease your vehicle, but the credit goes to the leasing company and not to you, unless the leasing company factors it into your monthly payment (which is unlikely, although being well-informed might help you negotiate for it).

As always, forgive me for focusing on only a single global market; I hope that my self-awareness of this habit serves as some sort of penance. I realize that incentives are available at the federal level in *many* countries throughout the world, such as Germany, the UK, and China. New Zealand has recently offered incentives for EV purchases as well as additional fees on vehicles that produce high levels of CO_2 emissions, which is exactly as it should be — New Zealand is a country that tends to do things the way they should be done.

- Navigating (or avoiding) the dealership
- Comparing buying versus leasing options
- Buying from Costco or other wholesale clubs
- Benefitting from corporate and credit card discounts
- Scoring incentives when leasing
- Shopping for a used EV
- Insuring your new wheels

Chapter 6
Buying Your New Electric Car

If you've already shopped around for an electric vehicle (EV) — by browsing online, renting for an extended test drive, taking a friend's EV out for a spin, or engaging in some combination thereof — and you have managed to narrow your selection to one or two models that you think would look *just right* in your driveway, it's time to pull the trigger and make the purchase.

Afterward, *you* can be that friend who pays it forward by letting others take your EV for a spin.

This chapter covers many of the considerations that impact the buying experience. As with so many other topics, knowledge is power — the power to make the best deal possible, the power to make sure that an outright purchase is right for your finances (maybe a lease is a better option), and the power to save when it's time to insure that car. (And you thought all the extra power available to EV owners comes from the battery pack.)

Let's start with a quick conversation about the car buying differences you may encounter when evaluating an internal combustion engine (ICE) car against one with an electric motor.

Exploring the Differences between Buying an ICE and an EV

Now, your first reaction to this section's title may be a question along these lines: "Differences? What differences?" If so, then

a) can't say that I blame you.

and

b) you're not alone.

In theory, the car buying experience should be the same, no matter which type of fuel and propulsion the vehicle uses. But as you can see throughout this chapter, you can find *significant* differences when shopping for and buying an EV as compared to the experience with an ICE car. Just as the electric vehicle has proven to be a disruptive force to the automobile as a product, it has also been disruptive to the product buying *experience*.

To see what I mean, let's start by taking a trip down to your local dealership.

Or, more to the point, let's explore the possibility of skipping that part altogether.

Navigating (or Avoiding) the Dealership

I'll begin with some word association. When I say *car dealership*, what other places come to mind in terms of your favorite places to pass an afternoon?

The principal's office? A dentist's chair? The DMV?

Yes, yes, everyone knows that dealerships are small businesses that provide jobs and offer important safety and maintenance services to the millions of vehicles on the road. The employees at dealerships support local communities by earning a paycheck, which in turn pays for roads, parks, schools, fire departments, and other community benefits. And the men and women who own the dealerships typically support economies by buying local radio and TV spots, sponsoring little league teams, (including my brother's — shout out to the Jimmy Dan Dodge Raiders!), and much more. They're pillars of the community, one and all.

The fact remains, however, that, for many people, visiting a car dealer ranks somewhere between shopping at a farmer's market and shopping at Costco, in terms of where they want to spend their Saturday afternoons. (Have you been to a Costco on a Saturday? If you value your sanity, pick another day.)

Yet that's where the cars are.

Or are they?

I promise not to go *too* far down this rabbit hole, but the car shopping experience has seen some *significant* changes that have, in my view, been pulled forward by the convergence of EVs and the dawn of the COVID era. For now, let's hit the Pause button on all this dealership talk as I ask you to consider one last complication.

Now then: Though it likely is *not* breaking news to the audience for this book, in the case of almost all EV-only manufacturers, visiting a dealership is not only unnecessary when making your car purchase — it's also not even an *option.*

Suppose that you've just finished reading your copy of *Electric Cars For Dummies* and, after taking a few minutes to leave a review on Amazon (how kind of you!), you decide that the very next thing on your to-do list is to purchase a new Rivian R1T pickup.

Oh, wait — you're not in Michigan, are you? Because if you are, buying a Rivian at a dealership in Michigan is a crime. And you can't test-drive a Lucid in Louisiana. Nor can you take delivery of your Tesla in Tennessee.

Weird, huh? Weird, but also true. As I write this, it isn't possible to have a Lucid, Rivian, or Tesla delivered to you if you're a resident of these states at the time of this writing:

- Alabama (no service center)
- Connecticut
- Louisiana (no service center)

- Michigan (no service center)
- Nebraska
- New Mexico (no service center)
- Oklahoma
- South Carolina (no service center)
- Texas
- Utah
- West Virginia
- Wisconsin

All three EV-only manufacturers mentioned in this section do offer mobile service, so the routine care and feeding of one of these cars can be performed — most of the time — while the car is in your garage (or driveway).

REMEMBER

Driving an EV helps reduce transportation emissions, obviously, yet a state like Connecticut won't let you buy an EV from an EV-only manufacturer. That's despite the fact that Connecticut, just like 14 other states, has adopted California's emission standards because they're more protective of public health and the environment than federal standards. Life is rich.

And the story goes even deeper than that. (I warned you that it's a rabbit hole.) For some reason that probably makes sense to the kind of people who yell at kids to get off their lawns, states like Colorado, Georgia (where Rivian is building a factory), New York, and others have regulations that limit the number of dealerships that can be opened by Tesla.

So then what's the deal — why can't I buy a Tesla in Texas?

The essence of the point-counterpoint when it comes to dealer franchises goes like this:

- **Point:** A car dealership you might walk into today while shopping for a Toyota or a Chevy has its origins in the 1920s or 1930s. At that time, American automakers — General Motors at first, followed shortly thereafter by Ford and Chrysler — began licensing the rights to sell their cars to independent dealers rather than sell directly to consumers.

  This business model generally worked well for both the auto manufacturer and the independent dealer because it freed up each one to concentrate on what it did best: Manufacturers focused on making cars while the dealers got to be viewed as pillars of the community — essentially, a car dealer's job revolves around building goodwill with the local community, to whom they eventually sell and service cars.

I say *generally* because the problem with this model was that manufacturers would occasionally establish their own licensed dealerships (among other abuses of power), giving them a huge competitive advantage over independent dealerships. A dealer that poured massive amounts of time and capital into the dealership could suddenly find itself in competition with a business owned by the very manufacturer that controlled the price of goods sold.

As a result of this naughtiness by the Big Three, dealer franchise laws were passed in all 50 states in order to protect independent dealers from unfair competition. By the end of the 1950s, any manufacturer selling directly to the customer was assumed to be engaging in an unfair business practice, and the dealership paradigm became as canonical as Din Djarin. (To find out what I'm referring to, check out *The Mandalorian* sometime.)

Until the one-and-only Elon Musk showed up and said, "Hold my beer" and then delivered this:

- **Counterpoint:** Tesla, unlike every other auto manufacturer operating in North America (at least I can't think of an exception), refuses to sell its cars to independent dealerships — full stop. When buying a Tesla (or now a Lucid or Rivian; I guess there's the exceptions), you're buying directly from the manufacturer, upending almost 100 years of car-selling tradition along with the laws in place to protect that tradition.

According to Musk, the direct-to-consumer route is simply a better deal for consumers:

"Existing franchise dealers have a fundamental conflict of interest between selling gasoline cars, which constitute the vast majority of their business, and selling the new technology of electric cars. It is impossible for them to explain the advantages of going electric without simultaneously undermining their traditional business. This would leave the electric car without a fair opportunity to make its case to an unfamiliar public." (See for yourself at `www.tesla.com/blog/tesla-approach-distributing-and-servicing-cars`.)

Is Musk correct? That's a conclusion I'll let you draw for yourself. What I can share, from both personal experience and several third-party anecdotes, is that walking into an EV-only showroom feels much more like visiting an Apple Store than a traditional car dealer.

If you're at the Cherry Creek Lucid Studio in Denver, Colorado, for example, you can expect to discuss only Lucid vehicles (and merch). You're likely to see only one or two Lucids there to sit in and/or test-drive. You may have to make an appointment for that test drive, similar to what many people do at the Apple Store when visiting the Genius Bar. They may also have a virtual reality setup for you to explore other options. If you decide to order, you won't haggle over price, nor will the consultant make you wait in a room alone while they supposedly confer with their sales manager over the suggested price.

In fact, and I'm telling you this because I work there, when you order a Lucid from a Lucid store (see Figure 6-1), you place the order on an iPad.

FIGURE 6-1: A Lucid store — not your father's car dealership.

Lucid Motors

Have I made the point yet that it's like the Apple Store? *Because it's like the Apple Store, people!* And because Tesla, Rivian, and Lucid stores tend to be in high-traffic areas like malls and shopping districts, an Apple Store may even *be* a few doors over.

Remembering that no states have banned EV car sales

Though it is true that you cannot buy a Tesla in Texas — the very state where Tesla is now headquartered — I want to be crystal-clear about this topic because it's easy to get confused: *No states in the US have banned electric vehicle sales.*

Yes, some actual humans in Amarillo own Teslas.

You absolutely *can* own one — or dozens! — of the aforementioned EVs in any of the nifty-fifty United States, and you indeed have to register and pay taxes on your EV(s) in the states mentioned earlier, in the section "Navigating (or Avoiding) the Dealership." The catch is that you can't actually make the vehicle purchase in states that have franchise car dealerships laws — in a physical location that exists to sell new automobiles.

So, what do you do? You buy online.

In the case of Teslas in Texas (it kind of rolls off the tongue, doesn't it?), residents of say, Dallas, would buy their car online and then have it shipped to a Tesla service center in Little Rock or Tulsa and *then* do the final paperwork vehicle pick-up there.

Why would Tesla move its headquarters to a state where the products it makes can't be sold to the state's residents? It has to do with another T-word: taxes, or more specifically not having to pay state income taxes on billions upon billions of capital gains wealth generated in another state. If you think it has anything to do with anything else, I'll see you at the next Flat Earth Society Meetup.

THE *FOR DUMMIES* CONNECTION

Some of the background info for this sidebar comes from a column from *CBS Sunday Morning* and *New York Times* correspondent David Pogue, who got his start in the mass media business by — yep, writing a *For Dummies* book! (Several, actually. David wrote *Macs For Dummies, Classical Music For Dummies,* and *Magic For Dummies* before originating the *Missing Manual* series and a career in newspapers and on television.) As they say, game recognizes game.

As a tech columnist, David covered the Tesla dealership kerfuffle back in 2018, when the list of states banning Tesla sales was even more onerous than the one shared earlier in this chapter.

If you're curious to know more about the changes since 2018 and use them to extrapolate probabilities for future changes in state dealership restrictions, I invite you to visit `www.yahoo.com/lifestyle/cant-buy-tesla-states-161318245.html`.

When I glanced at the 2018 map in David's piece, it hit me and I thought: "Oh, yeah." Jen and I picked up our 2018 Model 3 at a service center in Missouri. It was still illegal back then in Kansas.

In any event, the current laws around dealer franchises are an obstacle to EV adoption. They make it so that test-driving an electric car from certain manufacturers — the ones who only make electric cars — is off limits to millions across the United States.

REMEMBER

The point-counterpoint I discuss earlier in this chapter (in the section "So then what's the deal — why can't I buy a Tesla in Texas?") will continue in courts and state legislatures as reasonable people disagree about whether and how dealer franchise laws are applied in 21st century America. And, if you've been keeping up with the EV news lately, you may have noticed that states like New Hampshire and Maryland have already changed their state laws.

Of course, if you're *not* shopping directly from an EV manufacturer, one way or another you'll end up at the place where you likely purchased your last car: the independent dealership.

Buying an EV from a dealer

In theory, buying an EV from a dealership is exactly the same as buying an ICE vehicle. In other words, it doesn't matter to the dealer whether you're buying a Mustang Mach 1 or a Mustang Mach E — you'll still navigate the same intentionally complicated and confusing maze that's associated with traditional car buying. (Let's put a pin in that Mach 1 versus Mach E callout, though, shall we?)

As with any car purchase, be on the lookout for these potential problems:

- Confusing window stickers listing "special value packages" and similar items
- Useless add-on products, like undercoating, rustproofing, fabric protection, or theft deterrents
- Wheel and tire protection packages
- Bogus charges, like inspection fees or safety inspection fees
- Advertising fees, documentation fees
- Nitrogen tire inflation (when the air we breathe is already 80 percent nitrogen)
- Offers to "hold" the vehicle for a deposit
- Extended warranties
- Misleading information about leases and/or manipulating customers into leases

If anything, all these dealer scams (oh, sorry — *practices*) have only gotten worse in recent months as supply chain issues have cascaded into dealer inventory shortages, which in turn have led to all sorts of shenanigans from dealers, like the hand-scrawled "market adjustments" on the Monroney stickers shown in Figure 6-2. Some part of me believes that this image was Photoshopped, even though it wasn't. But hey, those courtside NBA seats don't just fall off of trees. (#hustleculture, amiright?)

RDeFran

John Voelcker

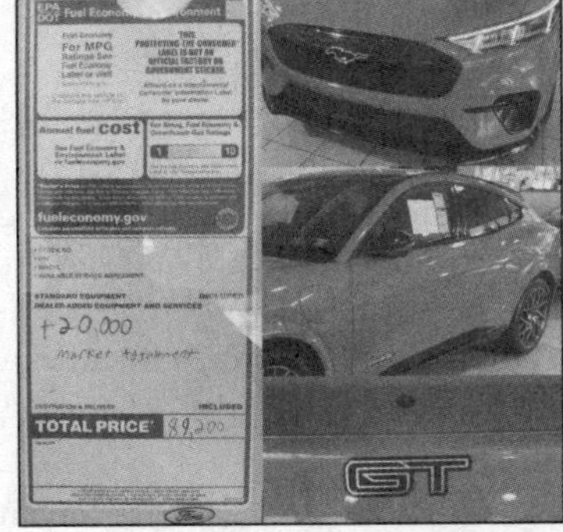

Tesla New York

FIGURE 6-2: At least the documentation fee now looks like a steal.

It makes a person want to go tire shopping at Costco instead, where they'll offer you — oh, no! — free nitrogen-inflated tires, in a ploy to get you to come back and have your tires serviced and replaced at Costco. Ugh. I warned you that Costco is depressing. (This topic has me wondering whether it's just a coincidence that all the car dealers in Fremont, California, are right next to the Costco there or a masterstroke of city zoning. Hmm.)

Anyway, these gotchas, and this advice, have been covered more extensively and much better elsewhere — specifically, in the excellent *Buying a Car For Dummies*, by Deanna Sclar.

Handling EV-specific issues at the dealer

Beyond the (sadly) typical landmines to avoid when visiting a dealership, especially in today's supply-constrained environment, EV buyers must also be prepared for the kinds of potential obstacles described in the following three sections.

Having to pay more for an EV in stock

Yep — you may end up paying more for an EV that's in stock, because EV manufacturing cannot keep up with demand, especially for the most anticipated models. (I'm looking at you, Ford F-150 Lightning, and you, Hyundai Ioniq 5, and you, Chevy Silverado E Sport.)

Be reassured, however, that many dealers, out of the goodness of their own hearts, have responded by assuring consumers that they won't try to leverage these shortages to their advantage.

In a parallel universe, that is.

In the universe you and I inhabit, unfortunately, dealers are trying to gouge the heck out of buyers who have lined up for vehicles where demand significantly outstrips supply.

That said, know that the practice is frowned on by manufacturers in the strongest possible way that manufacturers can frown on independent dealers (who, in theory, should be able to charge whatever price they want; it's that whole manufacturer-versus-franchisee issue once again).

Here are two recent pieces that chronicle the attempted practice, and what manufacturers are doing about it:

- `https://electrek.co/2022/01/21/gm-threatens-action-against-any-dealerships-attempting-markups-on-upcoming-evs`
- `https://electrek.co/2022/01/07/ford-warns-dealers-taking-advantage-f150-lightning-reservation-holders`

TIP

Do not pay a dealer markup for an EV. Never. Hit the secret Eject button (found in all EVs) to exit the vehicle, use the jetpack to flee from the premises, and then send an email to the EV manufacturer so that the dealer in question can be properly shamed. (Okay, I was kidding about the Eject button and the jetpack.)

Having to determine whether the dealer is even familiar with EVs

To be able to sell EVs, many manufacturers are *requiring* their independent dealers to become EV-certified. This process is not free, and some dealers have opted out. According to reporting from CNBC, roughly one out of every three Ford dealerships has said no-thank-you when presented with the opportunity to sell the Ford Mach-E and the F-150 Lightning. The reporting goes on to say that Colonial Ford, in Danbury, Connecticut, is "one of 2,300 Ford dealers, among a total of roughly 3,000, that have volunteered to become EV-certified, an investment that entails training sales and service personnel, upgrading battery charging stations, and purchasing special equipment, parts and tools. The remaining third have thus far opted out of spending nearly $50,000 for the certification. Other manufacturers are asking for upward of $300,000 for the designation." You can read the article at `www.cnbc.com/2021/06/13/gm-ford-are-all-in-on-evs-heres-how-dealers-feel-about-it-.html`.

One genuine concern for any dealer is the fact that EVs require no oil changes, transmission repairs, or other services that owners of ICE vehicles routinely bear — and such services account for 50 percent of dealers' gross profits.

REMEMBER

Though you may indeed be able to walk into any Ford dealer and buy a Mach E, you're likely signing up for future headaches when trying to get the vehicle serviced at the same dealership. If they opted out of the certification program, they may simply not be equipped for service.

Dealing with salespeople who may try to steer you away from an EV anyway

In a worst-case scenario, the salesperson you draw might simply harbor some kind of grudge against EVs.

For example, a 2019 blind shopper survey of 579 Sierra Club volunteers uncovered a rather troubling baseline about buying EVs from traditional dealers. Here are three highlights from the report:

- In 28 percent of dealerships visited, salespeople provided *no information* about how to charge an EV; in 31 percent of dealerships visited, salespeople provided no information about state and federal incentives.
- When volunteers asked to test-drive an EV, the vehicle was insufficiently charged — and was therefore unable to be driven — 10 percent of the time.
- Of the dealerships that sold EVs, more than 66 percent did not display them prominently, with vehicles sometimes buried far in the back.

Check out the full survey results at www.sierraclub.org/press-releases/2022/01/sierra-club-releases-first-ever-nationwide-investigation-electric-vehicle.

Surely the needle has moved by now, but still: In the more benign cases, it's probably a byproduct of the dealership not having the apple of your car-buying eye sitting on the lot (because of short supply), and not making as much on the service side of an EV after the sale anyway. (The salesperson doesn't necessarily even care about this aspect, of course, but their manager very well might steer them away from the *potential* close of an EV sale in favor of the *immediate* close of an ICE sale.)

A combination of these variables may result in your being asked to reconsider your EV-buying convictions. You may be asked whether you've truly "thought through" the whole charging issue. (You have — you're reading this book, after all.)

Mostly, though, this type of EV avoidance on the part of salespeople isn't as nefarious as either of the situations I've just described. Instead, it comes down to a matter of familiarity.

Car sales is a high-turnover game, yes, but the people there in the showroom are simply better trained for, and have more experience with, selling ICE cars. And people gravitate toward what they know. If sales staff were true believers in the coming EV revolution while surrounded by a lot full of ICE vehicles they're supposed to move, they likely would have headed for the exits by now.

So, despite the industry messaging about being "all-in on electrification by 2035" (looking at GM, which delivered 25 Bolts and 1 Hummer in Q4 2021) or about expecting "40% to 50% of its global vehicle volume to be fully electric by 2030" (um, tell me, Ford: Who will buy the other 60 percent of your cars?), today's dealership-bound EV buyer is working against years and years of inertia from the carmaker and dealer alike.

To wit: Ford's very own Explore Mustang web page doesn't mention the Mach E as one of those Mustang models. You can find the GT or the Shelby or the Mach 1, but I guess the Mustang Mach E isn't really a Mustang? I don't know — I just find it all very odd.

Choosing between buying and leasing

When it comes to buying versus leasing, there's really no difference between the ICE experience and the EV experience. When you buy, you own the vehicle. You can drive it as much as you want and modify it however you like.

If you're financing the car, the bank (or another financing entity) technically owns the vehicle until the loan has been paid off, but the car is still titled in your name.

In the case of a lease, you're effectively renting the car from the leasing company. This means you've entered into an agreement to register the car in your name, insure it, and pay the annual property tax due, for example.

In addition, in most lease agreements, you're agreeing to limit your use of the car to a specified mileage threshold — usually, between 10,000 or 15,000 miles per year, although it's the mileage during the *life* of the lease that counts. If you have a 3-year/36,000-mile lease, for example, it doesn't matter whether you drive 30,000 miles in year 1 and then 2,500 miles apiece in years 2 and 3. If, however, you drive 30,000 miles in all three years and turn the car back in with 90,000 miles on the odometer, you'll likely be stuck with a bill that will leave your accountant frowning. Lease agreements can charge as much as 30 cents per mile for anything over the contracted mileage allowance, so even going 5,000 miles over the allowance would leave you with an end-of-lease-bill of $1,500.

If this situation happens to you, the smartest financial decision is to just buy the car. Most leasing options let you buy the car you've been leasing at the end of the lease term.

In any event, these six factors are part of a lease:

- **The original purchase price:** The amount the leasing company paid for the vehicle. (This amount matters — a lot.)
- **The residual values:** The amount the leasing company estimates the car will be worth at the end of the lease term.
- **The difference, or *capitalized cost*:** The total amount you pay in rent for the car.
- **The interest rate:** Interest on the rent payments affects the total you pay.
- **The lease term:** The length of time you'll lease the car. This figure also affects monthly payments. Generally speaking, the longer the terms, the lower the monthly payments.
- **The mileage allowance:** This is the "gotcha" amount that can cause a lease to end in tears (for the lessee, that is, not the lessor, unless we're allowing for tears of joy).

There are also many free and fantastic lease calculators just a click or two away, including this one from financial website NerdWallet: `www.nerdwallet.com/article/loans/auto-loans/nerdwallet-lease-calculator`.

Leveraging leasing incentives

As with EV buying incentives, leasing discounts are constantly changing, so anything I put in writing here in terms of the bottom line would be obsolete by the time this book goes to the printer.

Instead, I can leave you with a list of resources where I've found fantastic leasing deals, and I highly recommend keeping these bookmarked in your favorite browser for the next time you go EV lease shopping:

- **Electrek:** The ultimate source of truth on the latest EV leasing deals is probably Electrek's Electric Vehicle Lease Guide page: `https://electrek.co/best-electric-vehicle-leases`. The page is continually updated to bring you the most recent incentives. As I was drafting this chapter using my biro and Moleskine, I noticed that the site had been updated just two days earlier.

  Further, this particular Electric Vehicle Lease Guide is navigable by manufacturer, so you can shop for lease deals only from Ford, Volkswagen, Chevy, Polestar, or any of several other prominent manufacturers, with a few notable exceptions, as I detail in just a bit.

  Green Car Reports: A little further down the rabbit hole of EV leasing options, you'll find Green Car Reports (`www.greencarreports.com`), a site that publishes news, reviews, and previews of electric and hybrid vehicles from all across the globe.

However, when you start your leasing journey within the pages mentioned here, you'll soon notice that many electric-only manufacturers aren't listed, including carmakers like Lucid, Rivian, and Tesla.

There are a couple of reasons for these omissions. One is that these relatively fledgling companies either don't yet offer leasing options or just don't offer anything in the way of incentives to lease. If you lease a Rivian R1T or a Tesla Model X, for example, you're looking at a relatively high monthly payment (when compared to other EV leasing options), and you will almost always come out ahead — financially speaking — by buying the vehicle and then selling it after three years if the new-car smell is what gets you going in the morning (YOLO and all that).

That said, it's still possible to lease from one of the carmakers mentioned previously. On the Tesla website, for example, all you have to do is configure your car and your options and then choose Lease from the payment page. (See Figure 6-3.) At the time of this writing, the default payment for a Model 3 was $548 per month, based on a $4,500 down payment for 36 months at a mileage cap of 10,000 miles per year. You can also easily customize the payment, increasing or decreasing the down payment amount to have an inverse effect on the monthly payment amount.

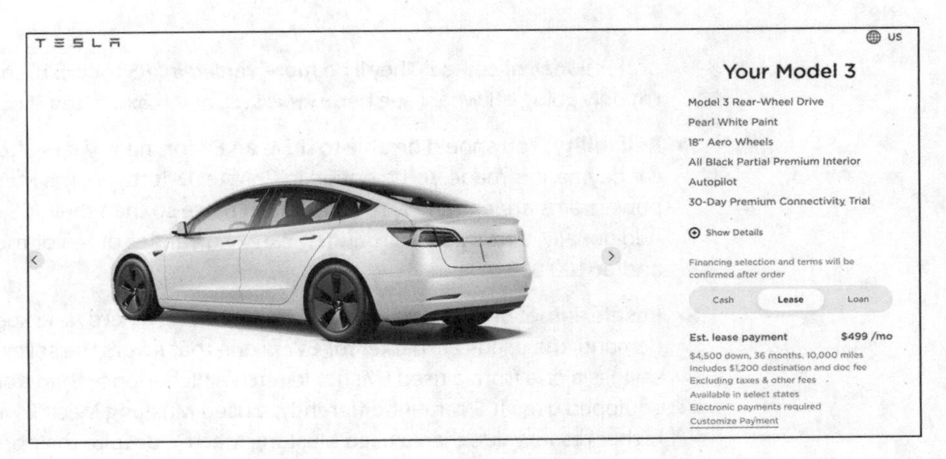

FIGURE 6-3: Leasing from the Tesla website.

Meanwhile, Lucid Motors launched its Lucid Financial Services (in partnership with Bank of America) in June of 2022, which allows customers to apply for 24-, 36- or 48- month leases, all without leaving the Lucid website.

The preceding list highlights the issues to consider when dealing with the lease-versus-buy decision. I leave some space later in this chapter for the debate over which option is better, but I'll provide my bottom-line advice right up front:

Buy an electric vehicle rather than lease it.

Taking advantage of the option to buy

Even though buying a new car has historically proven to be one of the worst uses of your money, here are three main reasons I've come to the conclusion that it's better to buy:

- **Flexibility:** It's your car, and you don't have to give it back to the lessor — ever. You can drive it for 2 months, 2 years, or 12 years. You can drive it 8,000 miles in a year or 8,000 miles on an epic road trip across the trans-Canadian highway and back. You can sell it any time during that period. In a lease, getting out early can be just as costly as a large overage in miles.
- **Tax incentives and write-offs:** Speaking of ownership, buying allows the buyer to claim any federal or state incentives. When you're not the buyer — in a lease, that is — those savings may or not be passed along to you.
- **Have a business? Buy the EV.** Depending on the nature of your work, the IRS *may* allow you to depreciate the full value of your new EV during its first year. Traditionally, leasing was the preferred way to go when deducting vehicle cost on a business tax return, but that's no longer the case. Check with your tax

TIP

professional, of course. They'll be more versed in IRS topic 510 than I am. I'm only going off what I see here: www.irs.gov/taxtopics/tc510.

- **Reliability:** You should be able to drive an EV for many years after the last car payment is made. As I mention in Chapter 9 (it covers maintenance), the powertrains are extremely reliable, much more so than their ICE counterparts. Additionally, battery packs routinely carry warranties of ten or more years and/or 100,000 miles.
- **Resale value:** Because manufacturers are having a hard time keeping up with demand, the used-car market for EVs is one that favors the seller. The gas savings alone from a used EV should fetch a higher price than a similarly equipped used ICE car. Put differently, a used Mustang Mach E will have a higher resale value than a used Mustang Mach 1, despite their being priced roughly the same when new.

In short, buying a new EV is a relatively low risk from a financial standpoint.

Despite all this, I do recognize that buying outright isn't the best idea for everyone, nor at every stage of their lives. I realize that sometimes a reliable newer car is worth more — in terms of safety, peace of mind, and professional appearances — than a paid-off beater, although this statement now almost exclusively applies to ICE cars because not many EVs are old enough to be classified as beaters.

Though I still prefer buying, mostly because of the residual-value equation, the good news — great news, really — is that you can find some amazing deals on EV leases.

If you know where to look.

Taking the route of Costco or another wholesale club

What is it about Costco that keeps me coming back for more? The $2 lunches? The rotisserie chicken? "Shopping" for a $3,000 massage chair for 10 minutes with lofi music playing in my headphones? Your guess is as good as mine.

Fine. As long as we're talking about Costco, I'd like for you to think of the first smell that hits your nose upon entering your local Costco. If you said the piping hot samples of microwave chicken cordon bleu, I don't believe you, because those samples are wa-a-ay in the back. You normally don't catch a whiff of those until you're checking off *saddlebags (2) of shredded cheese* and *post-apocalyptic mayo jar* from your grocery list, all the while likely thinking, "Why do I need all this, anyway? Am I shopping for a family or a zombie invasion?"

I digress. I'm talking, of course, about the *tires*. Costco sells and installs tires by the metric ton (or tonne, if you're in a Canadian Costco). And so it stands to reason that it also sells — or facilitates the sale of — the contraptions those tires attach to.

What I'm trying to say is that, by researching your Costco membership to see what deals might be available for an EV you're considering, you might uncover some truly significant savings.

For example, my research assistant (that would be me) just informed me that, as I was writing this chapter, *all* Chevy Bolt EV hatchbacks qualified for a $3,000 Costco member bonus. That bonus applied to both buying and leasing, and included the latest models, subject to availability.

In addition, my research assistant says that Costco memberships can be used to save on other EV models as well. Last year, Audi offered a discount on the E-Tron of up to $13,000 off the sticker price.

And I suppose now is a good time to share the Costco discount-finding methodologies of my intrepid assistant: He enters, or rather *I* enter, the words *"Costco EV incentives"* into a Google search and start clicking on the links that come up. He's a clever one, that assistant.

REMEMBER

As you've seen, Costco is a great place to purchase electric cars — however, there's no charging station in sight, at least not at any Costco I've ever been to, which makes visiting Costco even more depressing than it already is. People will wait in lines for 30 minutes to save $3 to $5 on a tank of gas. (See Figure 6-4.) And EV charging is what takes a long time? Huh.

FIGURE 6-4: "I only waited in line for 37 minutes to fill up my tank today" . . .said no EV owner ever.

Andriy Blokhin / Adobe Stock

My Costco membership card shows the worst picture of me in all of existence. It makes my driver's license photo look like it was taken by Annie Leibovitz.

As you can see, Costco and I have a rather complex relationship.

Taking advantage of a corporate discount

Did you know that if you work for National Grid, even part-time, you're eligible for a $5,000 discount on a Tesla Model 3? or that you can get a new 2020 Nissan Leaf, which retails for $31,000, for just $12,000? It's true. Check it out at `https://ngevcentral.com`.

So, what's National Grid, and how do you sign up?

Good question. Once again, I turn to my research assistant for more.

According to its website National Grid (`www.nationalgrid.com/us`) is "one of the largest investor-owned energy companies in the US." It serves more than 20 million people throughout New York, Massachusetts, and Rhode Island.

Well, that's not exactly *national*, is it? But neither is that the point. The reason a mention of National Grid graces these pages is that it offers a corporate incentive — a generous one, at that — to employees to help them save money on EVs. (A lesser amount is available for plug-in hybrid electric vehicles — PHEV cars, in other words.)

Okay, so, what if you don't work for National Grid and instead work for, say, the state of California? Over a quarter-million people do. And, if you're sitting in a coffee shop in Sacramento, there's a nonzero chance that you're also eligible for EV discounts not available to the general public. To see what you might be missing, check out the page at `www.dgs.ca.gov/PD/Resources/Page-Content/Procurement-Division-Resources-List-Folder/State-of-California-Green-Fleet-Employee-Pricing-Program`.

As you'll discover, through its Green Fleet Employee Pricing Program, California state and local government employees shopping for zero-emission or hybrid electric vehicles can receive discounts through contracted dealers. Similarly, customers of Southern Cal Edison may be eligible for significant rebates on new or used EVs. For more details, go to `www.sce.com/residential/electric-vehicles/ev-rebates-incentives`.

You get the idea. The landscape of EV incentives that are available to both employees and, at times, customers, is much too exhaustive to list in one place. My advice here is that it's worth a few minutes of alone time at your favorite company intranet or electric utility website to see what's up.

Shopping for a used EV

For most of my life, I've been interested in personal finance, and more than the average Brian. My minor back at Kansas State was in economics, and I've read dozens of books from authors like Peter Lynch, David Bach, Susie Orman, and many others. For better or worse, I could always explain the differences between fixed and variable rate annuities, and between Roth and SEP IRAs. I can have intelligent conversations about the virtues of a rising equity glide path, and I can tell you why you should ask whether your financial advisor is a fiduciary. (They should be.) In fact, I've helped several registered investment advisors become authors in their own right, having coauthored five books on financial planning.

What I'm getting at with all this is that the advice I've always heard, over and over, through all that reading and in all those interviews, is that when considering a car purchase, consider used as your first, second, and third options.

REMEMBER

Used cars are *always* the best financial choice.

I still agree with this advice in general, even though the "you'll always save buying used" axiom hasn't proven 100 percent reliable in my case. In fact, the most expensive car I've ever bought, in terms of a percentage of my net worth, was a used car. But more about that later. In any case, the first new car I bought in my life was the Tesla Model 3.

So, although the general advice of buying used remains sound, the caveat is that it *may* not be the case when it comes to an electric vehicle. It also very likely may still be — great advice, that is — but here's the thing: The world changed in 2020, and one of the things in that world was the market for used cars.

And I have the receipts to prove it. Figure 6-5 shows what my wife and I paid for our Tesla Model 3 in 2018.

Figure 6-6 shows what Kelly Blue Book says the car will sell for today, four full years later.

So, if I sold this car tomorrow, I'd be out-of-pocket about $3,000 over 48 months of ownership (excluding taxes and insurance, which do count, of course, but are not being calculated here out of nothing more than abiding sloth on my part), for an average monthly cost of $5 per month.

Whaaa? That's hard to beat from a total-cost perspective.

And yet, and yet, keep in mind that this is just one idiot's experience, and your mileage may vary. Significantly. I'm sure I could find several anecdotes about used car purchases from a few years ago where the seller actually *made* money on the resale. All it takes is a global pandemic, a supply crunch, and a lot of capital injected into world economies (justified, in my view), and suddenly everyone who bought anything pre-pandemic looks like Warren Buffet.

TESLA

Motor Vehicle Purchase Agreement

Vehicle Configuration

Customer

Order payment	$2,500
Order placed with electronically accepted terms	10/22/2018

Price indicated does not include taxes and governmental fees, which will be calculated as your delivery date nears. You will be responsible for these additional taxes and fees.

Description	Price (USD)
Model 3 Not including the $7,500 Federal Tax Credit	$35,000
Long Range All-Wheel Drive	$19,000
Pearl White Paint	$2,000
18" Aero Wheels	Included
All Black Premium Interior	Included
Subtotal	**$56,000**
Destination and Documentation Fee	$1,200
Total	**$57,200**

FIGURE 6-5: Hey! I guess it was $35,000 after all.

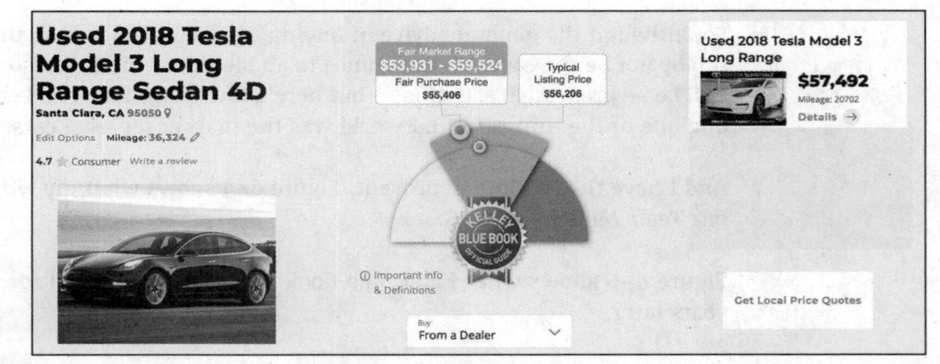

FIGURE 6-6: Numbers don't lie.

The bottom line is that *used* is still the soundest advice I can give.

But just like with any used car purchase, it's a lot more difficult to know exactly what you're getting. So, without further ado, here are a few unknowns that you should try to make known before pulling the trigger on that used EV:

» **How old is the battery?** Remember that the battery is the most expensive part of the EV. Though it's true that battery degradation isn't a big issue with recent-model BEVs, older EVs don't enjoy this luxury. If you're buying an older

Leaf, Prius, or BMW, it's worth double-checking the battery range as is. Don't rely on the seller to disclose this info.

- **How old are the tires?** Though the tire question is worth asking for any used car, heavier vehicles, like EVs, cause more wear on tires. That means a used EV with 40,000 miles can be a few thousand miles away from a pricey tire replacement — or it may have recently had that replacement.
- **Does the dealer (or the online retailer) have what you want?** One strategy I've used several times in my life to score great deals on used cars is to find something I like that there's lots of. For example: My last ICE purchase was a used 2014 Ford Fusion. Dealer lots at the time were lousy with off-lease Fusions, and when it came time to buy, I could choose between the black Titanium on *this* lot and the black Titanium on *that* lot, unless I changed my mind and went with the black Titanium on the *other* lot.

If I sound overly pleased with my 2016 car-negotiating self, trust your instincts. When it comes to playing multiple dealers off of one another in the quest to score the best deal on a used EV that everyone in town has in stock, good luck with that. The game, as they say, has changed.

Just to test it out, I've gone to Carvana and searched for a Ford Mach E. I see 18 results nearby, as shown in Figure 6-7. One is available for purchase. So, if I want a used Ford Mach E from Carvana, I have the choice that's variation on the phrase Henry Ford himself made famous 100 years ago: I can have any color I like, as long as it's white.

So, yeah. The biggest challenge in buying a used EV might be finding one you want in the first place.

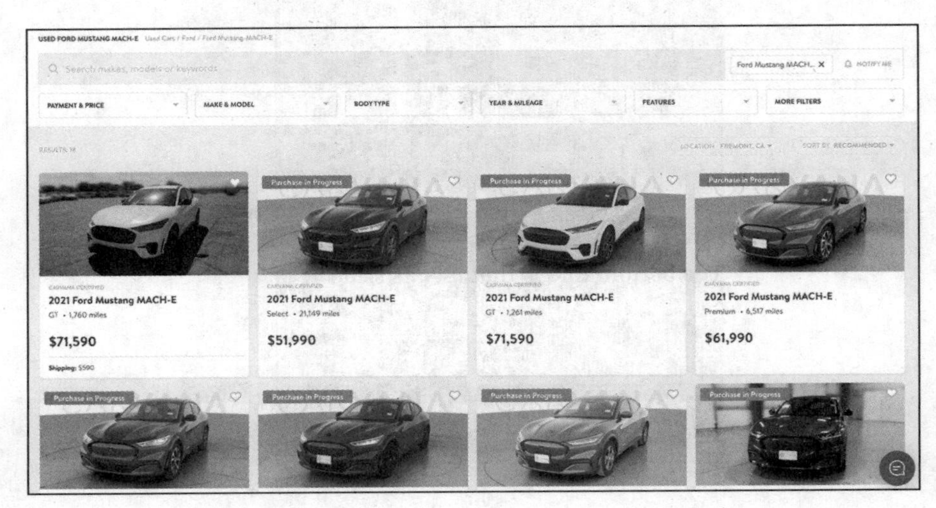

FIGURE 6-7: It's slim pickings when it comes to used EVs.

All of what I mention about used EV scarcity notwithstanding, there still remain some great options for shopping for used EVs. And the good news on this front is that there's no distinction between EV shopping and car shopping. Plug in the make and model for whatever suits your fancy on the website search forms and the world is your oyster. This list points you to some of the more popular options:

- **CarMax:** CarMax bills itself as the "nation's largest retailer of used cars," and I have no reason to think otherwise. As you'd expect from the largest used car retailer, it has plenty of EVs in stock, and it has a physical presence where you can touch, feel, and drive the very car you're considering before pulling the trigger. Or, as with Vroom and Carvana, you can keep it completely online.

  Plus, CarMax kind of pioneered the whole "the price is the price" part of used car selling that so many buyers find refreshing.

- **Carvana:** Carvana is an online retailer like, oh, I suppose Athleta or Lululemon, except instead of snug-fitting athleisure clothing, you can buy cars and have them shipped to your door. Its real claim to fame, however, lies in its car vending machines. (See Figure 6-8.)

  Car vending machines: kind of awesome — and also kind of weird.

FIGURE 6-8: Taking vending machines to a whole new level.

jetcityimage / Adobe Stock

As I discovered while shopping for the aforementioned Mach E, Carvana also lets customers call dibs on a car, just like you used to do with seats in the school cafeteria in sixth grade. It'll then hold the car for 30 minutes while you tick the necessary boxes to complete the purchase.

- **Vroom:** A competitor to Carvana in the online-only used car space, Vroom allows buyers to peruse "thousands of high-quality, low-mileage vehicles" in the browser tab next to the one you're likely using right now to search Google Images for car vending machines. So, while that tab is open (the Vroom one), know that it partners with major banks to help secure the financing part. In short, Vroom is a soup-to-nuts website that can help you find and buy your used EV.

In addition to these options, you can sometimes find used vehicles within the manufacturer's own website showcasing its new offerings.

For example, at the time of this writing, you can head over to `www.tesla.com` and peruse its used supply of Model S, Model X, and Model 3. (See Figure 6-9; no used Model Ys for sale on the day I checked.)

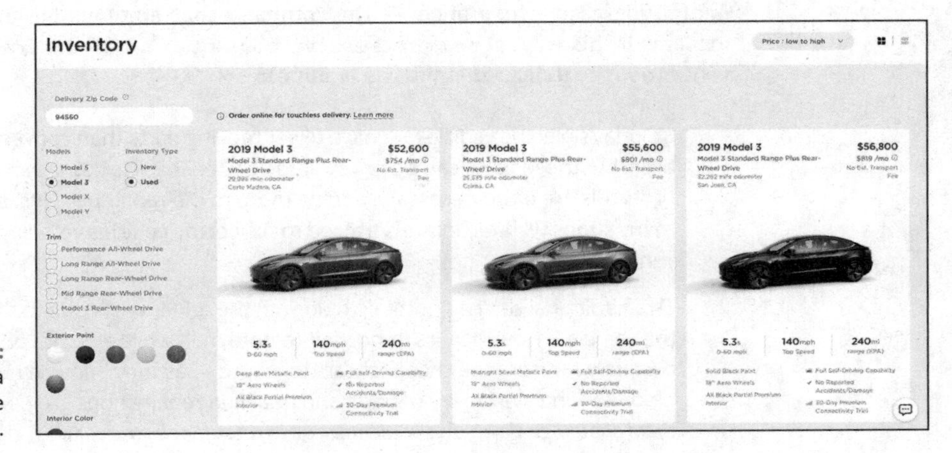

FIGURE 6-9: Searching for a used Tesla at the Tesla site.

Insuring Your Electric Wheels

Here's yet another topic that, were I reading this book in early 2018, before my first electric car purchase, I'd be tempted to skip. The 2018 version of me would have read the section heading and thought:

> "Insuring your EV? Yeah, I get how this works. Call your insurance company, give them the make and model of the car. Pay the difference in premium from your existing car (or perhaps get a refund), and that's that. What's there to think about? Next section, please."

As it so happens, Mr. 2018 self (along with you, reading this book in twenty-twenty-*something*), there can be some important differences between ICE and EV insurance shopping, and knowing about those differences can potentially save you quite a bit of money. What follows are a few best practices for EV insurance. (You can still skip the section, if you want; my 2018 self and you would get along famously.)

Let's start with the basics.

Seeing why EVs cost (a bit) more to insure

First of all, know that you'll almost always see a hike in your vehicle's insurance rates if the EV's purchase process is more than your current automobile. In other words, it simply costs more to insure a $55,000 Tesla Model Y than it does a $30,000 Nissan Rogue. Similarly, it costs less to insure a $55,000 Model Y than a $95,000 Porsche Cayenne e-Hybrid.

Secondly, EVs are generally more costly to repair, even for a similar type of accident. Progressive Insurance — the company that employs Flo and Dr. Rick — explains it this way, at `www.progressive.com/answers/car-insurance-electric-vehicles` (the italicized emphasis is mine):

> "[T]oday's electric vehicles . . . have fewer moving parts than conventional automobiles, but those parts can be pricey. If the battery pack is damaged, certain safety protocols are often necessary, adding more to the repair bill. Plus, there aren't as many shops with technicians trained to fix electric vehicles versus traditional vehicles.
>
> This additional risk has nothing to do with the driver — it's due to the technology in the car itself. However, it's important to note that, while electric vehicles are currently far from the cheapest cars to insure — as they become more commonplace and the availability of parts and qualified repair shops grows — *the cost to fix them should go down, as should insurance rates for electric cars.*"

In short, vehicle batteries are expensive, as are the technicians who are trained on how to replace them. But you knew that already, from having read this book. And, if you're reading this book two years or more after its initial printing, this is likely a moot issue because the roads are more and more filled with EVs.

And speaking of Progressive's mention of driver factors, as with all insurance, rates from even the same insurer for the same car can vary greatly, depending on the individual driver buying the policy.

For example, are you middle-aged with few to no traffic tickets or accidents over the past several years? Congratulations! Your insurance rates will be relatively low.

Are you still a teenager or over the age of 70? Hmm. I have bad news in the insurance premium department — according to stats gathered by the Insurance Institute for Highway Safety (IIHS), teens have *four times* as many crashes as drivers who are in their 20s. And though they get in accidents less frequently than all those damn whippersnappers, drivers over the age of 70 also have higher crash rates than people of a certain age.

REMEMBER

Teen drivers get into crashes because they have no concept of their own mortality. (At least that's my theory and I'm sticking to it.) Older drivers get into crashes because they're very much aware of their own mortality but don't particularly care anymore (again, my theory), and middle-aged drivers are both aware of their own mortality and gripped by a set of interconnected neuroses over a fear of said mortality, all of which causes them to drive more carefully (clearly self-evident to all with eyes to see).

"How to save 15 percent on . . ." (you know the rest)

Okay. So I've laid out the basics. But the main questions posed by this section remain: How do I save money on my electric car insurance? Is it even possible?

Fortunately, the answer is yes — if you know whom to ask.

Though it's true that most insurance companies do not offer discounts specifically for electric or hybrid vehicles, some are starting to come around. At the time of this writing, both Liberty Mutual and Travelers now offer discounts for EVs or hybrids.

Beyond getting a quote from one of these two insurers, what's the advice?

Well, you may have seen previously in this chapter that you may have limited options when trying to buy an EV, especially if that EV is used. But you have almost *unlimited* options about your insurance agent.

Call around! Get at least three quotes from the main players in the field, and make sure one of those is either Travelers or Liberty.

TIP

No matter which vehicle you own, EV or otherwise, always — *always* — review your uninsured motorist coverages when shopping for policies or even renewing one you already have. I can share from experience that when you're in an accident caused by an uninsured motorist, you won't be made whole by their insurance company, because uninsured motorists don't have one.

TESLA INSURANCE

When it comes to EV insurance, once again it's Tesla, that scrappy little trillion-or-so-dollar upstart, that is proving to be the outlier while also quickly establishing the new normal. Where possible by law, Tesla Insurance posits that the safer you drive, the less your premium will be, all other factors being equal. That's correct: Tesla offers insurance rates tied to driving *behavior.*

Now, if you've heard of offerings like the Progressive Snapshot or Geico DriveEasy programs, both of which use an app on your phone to track driving habits (there are others), you may be thinking that Tesla isn't the only entity taking this approach. Here's what Tesla has to say about the differences (emphasis mine):

> "Unlike other telematics or usage-based insurance products, Tesla does not require an additional device to be installed in your vehicle. Tesla uses specific features within the vehicle to evaluate your premium based on your actual driving. You will make monthly payments based on your driving behavior instead of traditional factors like credit, age, gender, claim history and driving records used by other insurance providers . . . [Y]our premium is determined based on what vehicle you drive, your provided address, how much you drive, what coverage you select, and the vehicle's monthly Safety Score. *An average driver could save between 20% to 40% and the safest drivers could save between 30% to 60%.*"

So, yeah. If you drive a Tesla, this is worth knowing about. And, if you haven't yet bought your EV and are crunching numbers to determine your true total cost of ownership, factoring in your monthly insurance premium is most certainly part of that algebra.

Insurance based on driving behavior is available in Arizona, Illinois, Ohio, and Texas. Yes, Texas, where you can't legally buy a Tesla but you can buy Tesla insurance. Sure, this makes sense.

You can learn more about that part of the equation at `www.tesla.com/insurance`.

Fun fact: Our teenage driver got into three fender benders, or at least three that we found out about. (Now that I think about it, there was very little fun about this fact. In 2019, we paid more to insure a 2001 Honda Civic than a 2018 Tesla Model 3 — hardly anyone's definition of fun.)

3

Maintaining Your Newfangled Horseless Carriage

IN THIS PART . . .

Master EV charging

Explore home charging options

Keep your vehicle's battery in peak condition

Take care of your vehicle's interior and exterior

» Exploring your options at home

» Exploring your options when traveling

» Treating range anxiety

Chapter 7
Charge 'Er Up!

Fueling up a gasoline-powered automobile only *seems* simple because you've spent your entire life doing it, or at least your entire life since you first got behind the wheel of a car. In that lifetime of experience of fueling up, you've usually had three options to consider when pulling up to the pump: regular, plus, and premium.

Can you guess how many options you need to consider when fueling up with electricity?

Yep. The answer is also three, and to keep things even simpler, the three levels of charging are referred to as — wait for it — Level 1, Level 2, and Level 3.

Here's where the gasoline analogy ends. Gas "levels" describe the octane rating — more or less the quality of the fuel. EV charging levels, on the other hand, describe how *fast* the fuel is added to the "tank." Electricity is electricity, after all, and really the only aspect worth considering for the EV driver is how and where that electricity comes from, which is a topic for another chapter.

This chapter provides the lowdown on those three EV charging levels: You'll see the speeds you can expect from each level; find out where to find, install, or use each of the three levels of charging; and even determine which EV fueling analogy is more helpful than trying to compare electric vehicle (EV) fueling to internal combustion engine (ICE) fueling.

I won't promise that if you read the entire chapter, you'll be ready to start your own electrician's business, but I can promise that you'll have much more confidence in determining what's available, what kind of charging level is best suited for your daily commute, and what you need in order to utilize each type.

Onions Have Layers, and Charging Has Levels

There was an animated ogre who once said to a hilarious donkey — you know what? I'm not even going to finish that sentence. If you saw the 2001 movie *Shrek*, in which Michael Meyers' ogre explained the emotional lives of terrifying bog monsters to Eddie Murphy's nonplussed talking donkey, you know the line. And if you have a daughter who was 3 years old at the time, you also have worn out a VCR tape (remember those?) of said movie along with the line ("Layers!") burned into long-term memory.

Thinking in terms of layers/levels sets the table for what's to come, so thank you, Shrek and Donkey, and let's get to it.

As you begin your journey with electric, you'll quickly learn — well, I guess you're learning *now* — that charging your vehicle means plugging in to one of three kinds of charging *levels*:

> Level 1, Level 2, Level 3

And, as you'll further see, these levels have implications about charging voltage, frequency, and current. However, the most significant aspect EV drivers will associate with charging levels has to do with *speed* — as in, "How rapidly will each level of charging put range back into my car's battery?"

You'll get a chance to explore each one, starting with the slowest level, Level 1. (Does this make Level 1 the onion's outermost layer? Or the inner . . . I'll stop now.)

Getting Up to Speed with Level 1 Charging

Generally speaking, the higher the charging level, the higher the power delivered by that charger, resulting in a faster charge rate for your EV.

At the *slowest* end of the charging spectrum is Level 1 charging. A Level 1 charger supplies power at a rate of between 1.3 kilowatts to 2.4 kilowatts, using (in the Americas) a 120-volt AC circuit.

When using a Level 1 charger, sometimes referred to as a *trickle* charger, you can expect to refill the battery at a rate of about 3 to 5 miles of range per hour of charge.

The good news about Level 1 charging is that you can find chargers almost everywhere. A Level 1 charger is simply a standard household wall outlet that provides power output at a rate of 120 volts. If you live in most of North, Central, or South America, the 120V outlet is part of your daily experience, as shown in Figure 7-1.

IcemanJ / Adobe Stock

FIGURE 7-1: A standard 120V outlet.

This thing you've been around your whole life has a name other than power socket, and that name is the National Electrical Manufacturers Association (NEMA) 15-5 outlet, which is a helpful thing to know if you're the kind of person who likes to have lengthy conversations with electricians. If that's you, you might as well know that NEMA also describes a standard for plugs, which in the case of the NEMA 15-5 plug, will also look familiar. (See Figure 7-2.)

NEMA 15-5 describes a 2-prong plug — the positive and negative pins — with a ground pin underneath. Didn't know those things had names, did you? Well, even if you did, *I* didn't know before I purchased an EV. Now I *do* know, and I think it's pretty cool.

FIGURE 7-2: A NEMA 15-5 plug.

Evan-Amos / Wikimedia Commons / Public Domain

Is Level 1 charging enough for your EV driving needs?

For most people, the answer to this question is yes. As I've mentioned several times, the average driver clocks about 30 to 35 miles of driving per day, which almost exactly matches what a Level 1 charger adds during an overnight charge: 30–50 miles of range. Fully charging a depleted battery, however, can take over 24 hours.

In my own experience, I can attest that during my first three months of owning an electric car, I used *only* Level 1 charging and it worked out just fine. It was 2018, but I can't recall ever using our family's gas-powered car (long since sold) over the fear of not having enough charge to arrive at our destination.

A typical week of driving began with about 270 miles of range, which was used for a 30-mile daily work commute, followed by evening trips to gyms, restaurants, movies, and rec league hockey games. We lived in a fairly typical Midwestern American city in terms of land usage — these activities and establishments weren't exactly cheek-to-jowl.

In any event, I plugged in the battery at night during the week, and usually finished our Friday evenings with between 50 to 100 miles of range. I then topped up the battery over the weekend, charging during times the car was otherwise idle in an office parking lot during weekday afternoons.

How do you use Level 1 charging?

Throughout most of North America and South America, using Level 1 charging couldn't be easier. Simply put, if you can plug in a lamp or toaster, you can plug in the car. All you need is your car's mobile charger and, if it's not attached already, the NEMA 15-5 adapter.

What are some obstacles to Level 1 charging?

The biggest challenge to Level 1 charging comes down to *access*. Yes, you can find Level 1 outlets everywhere — I'm writing this chapter in a very small room that can nonetheless charge six EVs simultaneously — but having Level 1 charging available *where you park your car* is another matter.

If you're parking on the street, under a covered parking awning at an apartment complex, or in a parking garage, for example, you probably don't have easy access to an outlet. In that case, it's good to consider where you might be able to access Level 2 and Level 3 infrastructure.

In my hometown of Fremont, California, it's a common sight to see a full row of Teslas parked at the superchargers at the Target store on a Sunday evening. These vehicles don't belong to people starting road trips at 6 P.M. as the weekend draws to a close. These are apartment dwellers who are concurrently grocery shopping and car fueling as they prepare for the week ahead. It's not the cheapest way of putting electrons in the battery pack, but it's still a lot less expensive than buying a tank of gasoline in California.

TECHNICAL STUFF

Tesla announced recently that they are no longer including the mobile charger as a standard item. Why? Apparently because their internal data showed that many Tesla drivers never used it. I have heard anecdotally that renters in particular face this issue, either plugging into charging units installed in an apartment parking garage or doing weekly charging at a Supercharger.

Come to think of it, we almost never use our mobile charger, either. We have a wall charger installed in our garage, and the mobile charger leaves the trunk about once a year.

Level 1 cheat sheet

The TL;DR (too long; didn't read) version of Level 1 charging goes like this:

Outlet: A standard household outlet (in most of the Americas)

Voltage/current: 120 volts, 60Hz AC

Connector type: NEMA

Power: 1.3–2.4 kilowatts per hour

Rate of charge: Four to five miles of vehicle range per hour

Cost to charge: $9.30 for 70 kilowatts (at 13.3 cents per kilowatt-hour, the US average price for electricity, charging from empty to full)

If you're the type of insufferable person (like me!) who points out information nuggets like it's the Beren*stain* Bears, not the Beren*stein* Bears, you'll delight in knowing that American appliances run on 110 volts of power, not 120 volts. Your car, however, can pull all 120 volts of juice from that outlet.

Turning Things Up with Level 2 Charging

Now let's — wait for it — *amp* up this EV charging overview by talking about Level 2 charging. (Boom! That's a top-shelf electric current terminology pun, which is to say that it's a lame one; the two aren't mutually exclusive.) As you might guess, Level 2 charging serves the same purpose as Level 1 charging — it just does so considerably *faster* than Level 1.

Level 2 charging supplies between 3 and 19 kilowatt-hours of power using a power circuit of 208–240 volts. This translates to adding 20–50 miles of range per hour of charging, beating Level 1 charging by a factor of 5 or more. In other words, Level 2 charging does in an hour what the Level 1 charger does overnight.

You can charge at Level 2 speeds while out and about, either running errands or at the office, or while parked in your (home or apartment) garage.

Level 2 charging at home

Level 2 home charging works just Level 1 — you plug the car in and let it charge.

In terms of the connectors used during a Level 2 charging session, we're typically talking about a NEMA 14-30, 14-50, or 6-50 outlet/plug combination (throughout most of North America and South America), although other outlets/plugs facilitate Level 2 as well.

Figure 7-3 shows what each of these looks like.

As with the 120V outlets, the good news is that most US households already have the infrastructure necessary for 240V circuits (which are created by combining two 120V circuits at the fuse box). In fact, if your home has a washer and dryer, chances are good that they're plugged into 240V outlets already, although they might not actually draw quite that much power.

To use the Level 2 home charging, just plug your mobile charging cable in and Bob's your uncle, even if you're not in the UK. Alternatively, some homeowners (and apartment landlords) are installing dedicated wall chargers in their garages so you don't have to muck about with your mobile charger. I'll discuss that option later in the chapter.

FIGURE 7-3: NEMA 14–50 and NEMA 14–60 outlet/plug combinations.

Enphase Energy, Inc.

TIP

Using an Airbnb or a VRBO and have an EV? Check the listing to see if it's got a 240v outlet.

TECHNICAL STUFF

Why did I mention 208 volts? Because for reasons that are beyond my limited understanding, commercial properties typically use 3-phase power with 208 volts. (3-phase power refers to the use of three wires to deliver the electrical load, single phase refers to a single wire circuit. Residential homes are almost always served by a single-phase power supply. If you look at a standard household outlet, you're seeing a port for the single-phase wire (the "hot" wire), a port for the neutral wire, and a port for the ground. Three-phase power can deliver three times as much power as a single-phase supply.)

Level 2 charging in public

And the good news continues. Level 2 is by far the most frequently installed type of EV charger in the public space, having popped up in recent years in parking garages, grocery stores, restaurants, hotels, shopping centers, movie theaters, and other locations like the workplace.

Office buildings are beginning to make Level 2 chargers standard issue, and workplace charging is offered as a perk from many employers. Yes, I work at an EV manufacturer, but we get 10 kilowatt-hours for free before the meter starts running — if we arrive early enough to claim a spot, that is.

Figure 7-4 shows where the cool kids park when reporting to the office.

FIGURE 7-4: My home (Level 2 charging) away from home (where I have Level 2 charging).

TIP

Because of the wiring needs and the installation complexity involved (construction permits and such), Level 2 chargers are usually located close to the building in office and shopping complexes. Rock star parking for you! (I guess you can get your daily steps in by taking the stairs.)

You may also see Tesla's destination chargers bolted to a wall at a hotel or a post at a favorite restaurant. These, too, are Level 2 chargers, and they're *extremely* common at popular California destinations. Wineries, hotels, restaurants . . . you name it. I've lost track of how many times I've pulled in somewhere and thought, *cool, they've got a destination charger here*. They've even got a couple installed at the Grinders Stonewall in Lenexa, Kansas. (Hello, Grinders. I miss you; please ship me a pizza when you get a sec.) Figure 7-5 shows what the destination charger looks like.

And here's the thing: Even though the destination charger is Tesla-branded and is fitted with a plug that works with Tesla vehicles, other EVs can plug in as well. All it takes is an adaptor.

Speaking of plugs and adaptors, it's also very helpful to understand that, in the US, Tesla uses a proprietary plug when charging. Figure 7-6 shows what it looks like.

FIGURE 7-5: Luckily for you it's not your Final Destination.

Mediaparts / Adobe Stock

FIGURE 7-6: Tesla's charging connectors in the US are specific to Teslas. Kind of analogous to Apple's Lightning ports.

Mak / Adobe Stock

Meanwhile, most other EVs use a more standardized plug and connector for Level 2 charging called the J1772, which sounds like the name of a droid in a Star Wars prequel but is in fact related to EV charging. (Someone needs to hire a branding team.) It looks like what you see in Figure 7-7.

TIP

If your work doesn't offer charging, ask them about it today. A fellow colleague most likely has the same question. Many governments and utilities offer tax credits, grants, rebates, and other incentives to encourage workplace charging installation.

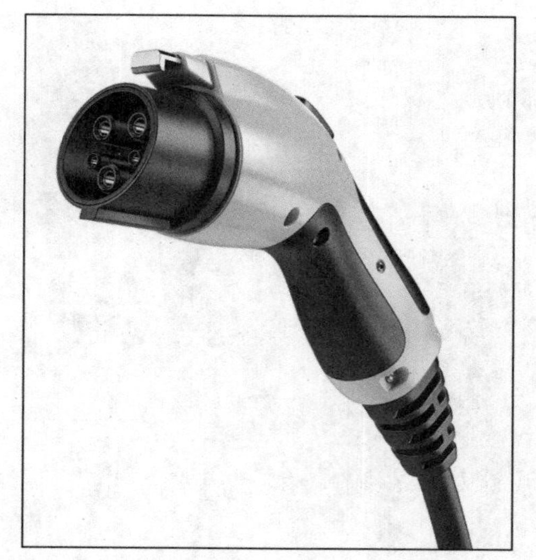

FIGURE 7-7: If it's not a Tesla, it will use this connector for L2 charging.

Sashkin / Adobe Stock

Not *all* public Level 2 chargers and electric vehicles are compatible with all EVs, but this should be a very rare exception. The rule, as in a rule that applies in 99% of the cases you'll use Level 2 public charging, is that if you find a Level 2 charger, you can charge your car, assuming you have the J1772-Tesla plug adapter handy in your glovebox.

The main takeaway for the EV owner is that when using Level 2 charging, you're adding more miles of range in a shorter time when compared to Level 1 — again, it's somewhere in the range of 20 to 50 miles per hour.

Should you install a Level 2 outlet at your home?

Yes.

Unless it's simply cost-prohibitive, knowing that you could charge your car from empty to full overnight is more than worth the initial setup expense during the first few months of ownership. Over the anticipated 6- to 8-year lifespan of the vehicle, it will more than repay you in terms of time saved.

Houses, condos, and apartment complexes without the Level 2 charging infrastructure will be at a significant disadvantage in the marketplace. So one way to frame it is that you'll either pay now or pay more later on.

TIP

My advice on home Level 2 charging is to just write the check, get the outlet installed, and rest easy at night knowing that your EV's battery will be topped up every morning.

How do you use Level 2 charging?

Depends on whether you're using public or home/apartment Level 2 charging.

Using public L2 charging

As I mention earlier in this chapter, public Level 2 charging comes down to finding an available Level 2 charger and then plugging in your car, usually after signing up for an account through your smartphone so you can be billed for the fuel. The most ubiquitous chargers you'll find right now are from companies like ChargePoint, EVGo, and Blink.

These charging stations are frequently located in the parking lots of big box stores, shopping malls, and parking garages. Level 2 public charging stations include the connector, so while you may need an adaptor, you can leave the mobile charger in the frunk.

Using home L2 charging

For home use of Level 2 charging, just plug the mobile connecter into the wall outlet. And you needn't worry about accidentally plugging in your 120V charging adapter to a 240V circuit, just like you needn't worry about plugging in your electric dryer into a standard wall outlet: the plugs are incompatible.

You can also use a dedicated wall connector. Many EV manufacturers make their own, dedicated Level 2 chargers, which are typically installed by a certified electrician and connected to either a 240V circuit in a home or a 208V circuit at a commercial property (hotels and office buildings, for example). Figure 7-8 shows you what I'm talking about.

Besides the auto manufacturers themselves, lots of third-party companies also make EV wall chargers for home or commercial use.

Some of these charging stations offer *smart* features, such as the ability to be controlled over Wi-Fi or Bluetooth using your phone, or even with Alexa or Google Assistant voice commands. Though being smart won't make these devices charge any faster, they do allow for data collection, reporting, easy scheduling, and other tasks.

FIGURE 7-8: A Level 2 home wall charger.

Lucid Motors

TIP

The wall chargers from Tesla, Lucid, and Rivian also include such smart features. What's more, most of the manufacturer apps allow for scheduled charging — you don't *need* a smart charger, or *any* wall charger for that matter, to tell your Tesla to start charging (via the mobile connector) at 1 A.M. for example.

TIP

You *may* be able to get the cost of home L2 charging installation covered. As this goes to press, Chevy is offering to cover standard installation of a Level 2 charging outlet for customers who purchase or lease a Bolt EUV or Bolt EV. Be sure to check the manufacturer website before making the purchase, or perhaps bring up the Chevy offer in the course of negotiations.

Not every EV manufacturer also makes their own dedicated wall chargers, though. Fortunately, there are several companies who do. If you plan to have an EV parked in your garage for years to come, it might be a worthwhile upgrade to install dedicated wall chargers that can accommodate almost any kind of EV you'd want to park there.

In addition, these third-party dedicated wall chargers will be the ones most apartment management companies call when setting up charging infrastructure for their residents.

The following list, then, describes some fantastic Level 2 home charger options, along with some notable benefits offered by each one, according to the manufacturer descriptions. Thanks to online ordering, all these can be shipped your house in just a few days. (How fast they can be installed is another matter. It can depend on a wide range of factors, such as whether your electric service needs to be upgraded, how far the outlet is from your breaker box, whether you need an electrician or are a DIY type, and so on.)

- **ChargePoint Home Flex:** Up to 37 miles of range per hour using up to 50 amps of power. (Most drivers use 32 or 40 amps.) It lets you charge faster and smarter using the ChargePoint app, by letting you set a schedule to charge when electricity is cheapest. The app also lets you set reminders so that you never forget to plug in. To top it all off, it's Alexa compatible.
- **Enel X JuiceBox 40:** A 40-amp EV charging station for indoor/outdoor installations. Wi-Fi connectivity allows for remote control, scheduling, data monitoring, and charging reminders. It's also Alexa compatible.
- **Grizzl-E Level 2:** Uses a NEMA 14–50 plug with a 24-foot cable. Made in Canada — the Great White North — it's heavy-duty as well as suitable for both normal and cold temperatures. The Grizzl-E is compatible with all EVs and PHEVs sold in North America.
- **EvoCharge EVSE:** Fully compatible with all EVs and PHEVs sold in the United States and Canada, the EvoCharge EVSE has charging rates of between 25–35 miles of range per hour. It uses a NEMA 6-50 plug and a universal mounting bracket and is rated for indoor and outdoor use in all weather.
- **Wallbox Pulsar Plus:** A 40-amp charging station that lets you use the myWallbox app to wirelessly control and monitor your charger via Wi-Fi or Bluetooth. That means you can create schedules, receive notifications, set reminders, and more.

Honestly, you can't go wrong when it comes to these dedicated Level 2 chargers, whether shopping for home, office, or apartment complex. They all do more or less the same thing, look more or less like what you see in Figure 7-9, and, assuming proper installation, do their thing very well indeed.

TIP

These dedicated wall chargers are sold separately — usually, in the ballpark of $400 to $800.

What are some obstacles to Level 2 charging?

Seeing as you can't (normally) park your EV in your laundry room, the issues of outlet *availability* that apply to Level 1 charging also apply here.

What I will add, though, is that Level 2 charging infrastructure — especially the garage 240V outlet — is becoming more and more common, especially at new construction projects. For example, when searching for a place to live in California, all the places our family considered — both relatively new construction condo units and newish-construction apartment buildings — included either 240V outlets in the garage or had Level 2 charging stations (or wall connectors) in the parking structure reserved for resident use.

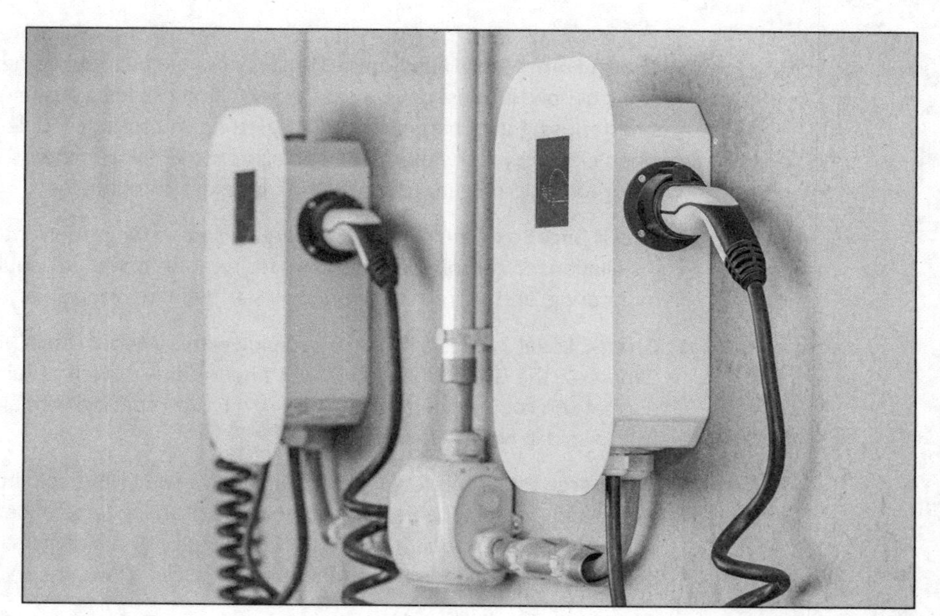

FIGURE 7-9: A dedicated Level 2 home/ apartment charger.

Kara / Adobe Stock

WARNING

However: Even if you *are* the property owner and decide to install a 240V outlet in/close to where you park your car, you can potentially run into issues like these:

- Your house's electrical service doesn't support adding another 240V outlet.
- Installing the outlet will result in minor (or major) demolition work in order to access conduit space.
- The fuse box and outlet location are far apart.

All these issues can be mitigated, of course, using the same solution that mitigates almost of any of life's transient issues: money. Depending on local market rates, existing electrical service, and physical space, some 240V outlet installations will run you $100, and others will set you back $10,000. In other words, electricians can usually install new outlets wherever you point, but sometimes the spot you're pointing at requires some minor (or major) demolition work. These same folks who have skill with electrical wiring are almost never able or willing or qualified to remodel drywall.

TIP

The *caveat emptor* to all this discussion about charging obstacles is that if your heart is set on Level 2 charging in your home, it's worth consulting an electrician *before* pulling the trigger on the purchase and then pulling into your garage for the first time.

But *should* you invest in a dedicated wall charger?

I think the answer to this one is "It depends." Yes, the mobile charger is functionally the same as a dedicated unit in terms of getting electrons from the 240V outlet to the car's battery. Also, yes, there are advantages to investing in a dedicated wall charger when getting that first EV.

I think, then, it mostly comes down to matters of a) aesthetics and b) convenience.

In terms of aesthetics, consider whether you're the kind of person who takes pride in the appearance of your garage. Is everything neatly organized on shelves? Are bicycles hung from the ceiling in order to maximize storage space? Is the floor epoxied? If so, a dedicated wall unit is probably right up your alley.

In terms of convenience, you should factor in how likely you are to remember to grab the mobile charger and put it in your frunk every time you need it on a road trip. Again, the advice here is to take the old Greek aphorism to heart: Know thyself.

Oh, and a third consideration is commute length. If you have a 50-mile one-way commute to work every day, then having a dedicated Level 2 charger in your home, apartment, or workplace is pretty much a need-to-have rather than a nice-to-have. Again, a mobile charger will work fine, but a dedicated wall charger just makes your life that much easier when you arrive.

Now, you *could* DC fast charge (discussed in the following section) for all your charging needs in such a long-commute scenario, but I'm not sure the time/cost involved would make it worthwhile as a daily habit. For the most part, you want your car charging where you work or sleep. Not only are you charging while the car isn't being used, but you're also getting the best rates on that electricity.

REMEMBER

Most modern electric vehicles can accept at least 32 amps of charge. The Chevy Bolt accepts 40 amps from Level 2 chargers, and both the Tesla Model 3 and Model Y are capable of 48 amps if using the Tesla wall charger.

Level 2 cheat sheet

In case you have an attention span conditioned by years of smartphone use, here are the essentials for Level 2 charging:

Outlet: 240V household outlet (in most of the Americas)

Voltage/current: 208–240 volts, 60Hz AC

Connector type: NEMA

Power: 3–19 kilowatts per hour

Rate of charge: 18–40 miles of vehicle range per hour.

Cost: Varies. If you're charging at home, the average cost is 13.3 cents per kilowatt-hour in the US, or from empty to full for roughly $10. Public stations charge roughly 30 cents per kilowatt-hour, so from empty to full for most battery packs is in the $20–$25 ballpark. Note that some employers or businesses now offer limited charging for free.

Getting a Wicked-Fast Charge with Level 3 Charging

Perhaps the most recognizable, most often discussed level of EV charging is Level 3. Ironically, it's also the least often used. (I don't have data on this; it's just an educated guess.)

In my experience, whenever someone curious (or skeptical) about EVs asks, "How long does it take to charge?" what they're *truly* asking about is the overall experience while using a Level 3 charger. The reason is that they're using an old fueling paradigm, something that I get to later in this chapter. (See the sidebar labeled "The yellow fuel light and the paradigm shift.")

In fact, you might have heard Level 3 charging discussed elsewhere as DC fast charging (DCFC). Or just "fast charging." These three terms are interchangeable and will be used that way in this section. Oh, and the reason it's called DC fast charging is yet another topic I'll address in a bit.

And now let's throw yet another synonym into the mix: because Tesla dominates both the mind- and marketspace for DC fast charging infrastructure, you may have heard it called *supercharging*. If you see the word supercharger, it refers only to something specific to Tesla.

Got it? Level 3, DCFC, fast charging, supercharging — these all refer to the same thing, which is this:

Much higher charging speeds than either Level 1 or Level 2. Level 3 charging supplies between 75 and 250 kilowatts of output per hour, meaning that it can add between 3 and 20 miles of range every *minute*. Figure 7-10 shows you what I mean.

In practical terms, a good Level 3 charger can take an EV battery from almost depleted to 80 percent in roughly 15 to 25 minutes.

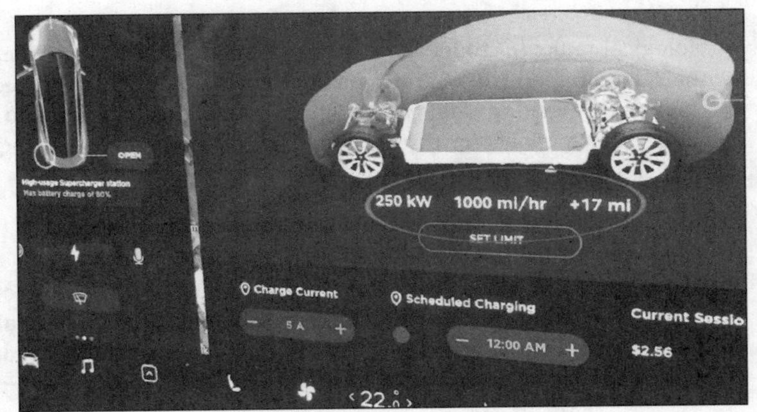

FIGURE 7-10: A Tesla, charging at 1,000 miles per hour.

Tesla Raj

So, as you can see, this charging is most targeted and appropriate for those folks going on road trips — DCFC is analogous to the neighborhood or roadside filling station. You can see from Figure 7-11 that the chargers themselves are even designed to resemble gasoline pumps. (It's all about that paradigm, yo.)

FIGURE 7-11: Going for a familiar look. Skeuomorph fans, rejoice.

ehrlif / Adobe Stock

REMEMBER

Again, using personal experience as a guide, I can report that using a Level 3 charger on a road trip is similar to the experience at Kwik Shop or Flying J or any other highway pit stop. In fact, this charger is often located at these very same gas/convenience stations. In the time it takes two travelers to empty the old biological tanks (or to let the dog empty his; what kind of barbarian doesn't take their dog on road trips?), refill on snacks and sundries, and check in with whatever needs checking, said two travelers have added between 200 and 250 miles of range, which allows for three more hours of driving, even at highway speeds.

Fast charging plug connectors

Fast charging is called *DC* fast charging because it utilizes direct current instead of the alternating current (AC) that's used in household electrical circuits. What's more, using DC rather than AC means that you might also encounter a different type of connector when using Level 3 versus Level 1 or 2.

REMEMBER

Batteries use DC, incidentally. Electric Vehicles have converters in the car circuitry that change AC power to DC for the battery consumption. One of the reasons why DC chargers can charge faster is that they convert AC power to DC at the charging station rather than in-car.

Let's break this down.

As I mention earlier in this chapter, unless we're dealing with a Tesla, Level 2 charging uses a standardized J1772 connector.

DC fast charging, however, can use one of several connector/plug combinations. The three that you're most likely to encounter are:

- **SAE Combo:** This Society of Automotive Engineers International (SAE) connector is known more commonly as the Combined Charging Standard (CCS) 1 in the U.S., and CCS2 in Europe.
- **CHAdeMO:** If purchasing an EV in Japan, you might end up using a CHAdeMo connector. CHAdeMO is the trade name for a DC fast charging system formed by the Tokyo Electric Power Company and five major Japanese automakers.
- **GB/T:** The GB/T standard refers to a combination AC/DC connector/plug commonly used in China.
- **Tesla**: Tesla uses a proprietary plug and connector. The advantage of the Tesla connector is that the plug is the same whether using Level 2 or Level 3 charging (and Level 1 for that matter). The disadvantage of the Tesla adapter is, well, that it's proprietary. So, adaptors.

Figure 7-12 shows what these different connectors look like. (Note that the Mennekes connector shown is just the AC part of a CCS 2 connector.)

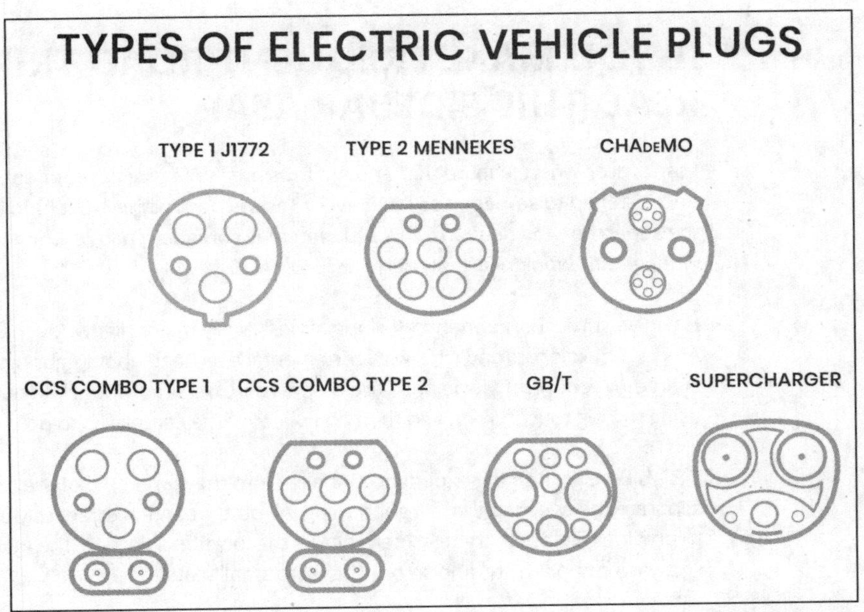

FIGURE 7-12: They call them visual aids for a reason.

svetolk / Adobe Stock

You may also note here that the CCS port is a *combination* of a J1772 AC connector and a two-prong DC connector. In fact, Teslas in Europe will come with a CCS 2 port under the charging door rather than the Tesla proprietary port deployed in US models.

So, do you have to worry about all these different connectors? Not really, although some cars aren't capable of fast charging. I'm thinking here of an older Nissan Leaf, or any Toyota Prius — you can't fast charge any plug-in hybrids at it currently stands.

That said, almost all late-model BEVs are equipped for DC fast charging — just make sure to take note of your car's port before you plug in.

REMEMBER

If you're without access to Level 1 or Level 2 charging, such as at an apartment complex that has no access, Level 3 charging might be the primary way to "charge 'er up."

Examining Level 3 charging claims from EV manufacturers

The question "How long does it take to charge this thing?" is such a prevalent question/obstacle for EV manufacturers that their boasts about charging speeds have been front and center in their marketing campaigns almost from the word *go.*

IF YOU MAKE FREQUENT ROAD TRIPS, READ THIS SIDEBAR ASAP

I can't recall where I heard this, and am thus unable to give appropriate credit, but I once listened to someone say that every Tesla is born with a trust fund in the form of the supercharging network. This situation may someday change, but at the time of this writing, I still wholeheartedly agree.

Having a supercharger network along almost every major highway across North America makes for road trips with zero mental overhead about figuring out where to charge. Whenever my wife and I are on the road, we simply punch in our destination using the big touchscreen and then do whatever the car tells us to do.

Tesla's Level 3 charging experience is baked into the ownership of the car. You cannot separate the two, and that's a good thing. All other manufacturers rely on third parties to provide the fast charging experience — the location selection, the maintenance, the software compatibility, and so on. This is no small feat.

What's more, ramping up Level 3 chargers is one of those chicken-or-the-egg conundrums for other manufacturers. For example, companies like Ford, Rivian, or Lucid have little incentive to spend their resources on a profitable charging infrastructure when Electrify America (the only other major builder of fast charging infrastructure in the US) is being forced to do it anyway. (Electrify America is the company Volkswagen was forced to spin up as part of their diesel-gate settlement with the EPA.) Electrify America, on the other hand, has little incentive to build out stations until a lot more Fords, Rivians, and Lucids are on the road charge at Electrify America (EA) stations so that EA can see a return on investment.

Tesla's supercharger efforts predated the build-out of Electrify America (and others). Tesla had little choice in the matter, and now, whether it was superior strategy or just blind luck, the supercharger network is certainly a competitive advantage.

So although this advice may not age well, my two cents is that if you're in, say, Kansas City and will be hauling kids to soccer tournaments in Wichita, St. Louis, or Des Moines every other weekend, Tesla is the BEV for you and your offspring.

As you'll see in just a moment, most of these claims more or less hover around the median charging rate (in terms of miles of range per hour) and are, in my view, negligible in terms of the everyday use of the car.

That said, it's worthwhile to examine some of the fast charging claims side-by-side. I'll start with the market leader.

TECHNICAL STUFF

My methodology, by the way, was entering the term *car charging speed* into a search engine and then looking for a link to the manufacturer website. Fancy, I know.

Here are the results:

- **Tesla,** `www.tesla.com/supercharger`: It has more than 30,000 global superchargers, it has a 250-kilowatt maximum rate, and it takes 15 minutes to add more than 200 miles of range.
- **Lucid,** `www.lucidmotors.com/air`: Charging specs are prominently featured, touting more than 2,400 chargers (by Electrify America), a 900V charging system, 300 miles of range in 20 minutes, and the capability for bidirectional charging (the car can charge your house).
- **Rivian** `https://rivian.com/experience/charging`: More than 3,500 DC fast chargers are being built exclusively for Rivian owners by the end of 2023. It features up to 140 miles of range in 20 minutes and charging power for over 200 kilowatts initially and more than 300 kilowatts in the future.
- **GM** `www.chevrolet.com/electric/bolt-ev`: Nothing appears on the Bolt EV information page about Level 3 charging, other than a mention of the MyChevy app, which can help you find Level 3 charging stations.
- **Ford** `www.ford.com/mustang/ev-charging/mache`: The site offers a few mentions regarding access to the Blue Oval charge network, though there's a lot of fine print to sift through. Though this is "North America's largest public charging network," according to Ford, that statement simply isn't true. Several footnotes, caveats, and asterisks make it *technically* true, but Blue Oval means that you can charge at Electrify America and EV Go — there's no such thing as a Blue Oval charger.
- **Volkswagen/Porsche** `www.porsche.com/usa/models/taycan/taycan-models/taycan-turbo`, `www.vw.com/en/models/id-4`: It offers 800-volt architecture. Otherwise, it's just the classic ad copy: "Feature, not a benefit." The Porsche website assumes that you consider 800V to be a big deal when it comes to charging — though it isn't, at least not to an EV owner. What matters is how many miles of range are put into the battery, and how quickly. The Volkswagen ID.4 website mentions access to Electrify America, but nothing about charging speeds.

The larger point is that EV manufacturers want you to be able to charge in a flash, and they have spent millions of R&D and marketing dollars to make it so. (It's also, well, *interesting*, that the BEV-only manufacturers have taken pains to feature their charging information much more clearly and prominently than the ICE carmakers.)

Identifying the players in the DC fast-charging business

As this section makes clear, many auto manufacturers not named Tesla are counting on the efforts of DC fast charging manufacturers to facilitate Level 3 charging (and thus facilitate more rapid EV adoption) along the planet's highways and byways.

These brands are leading the <ahem> charge:

- **Electrify America:** www.electrifyamerica.com
- **EV Go:** www.evgo.com
- **ChargePoint:** www.chargepoint.com
- **Blink Charging:** https://blinkcharging.com
- **SemaConnect:** https://semaconnect.com

Note that many of these Level 3 players are also prominent in the public Level 2 space as mentioned earlier.

Plus, Tesla is now starting to open up its supercharging network to non-Tesla vehicles. A pilot program has just been implemented in the Netherlands as this book goes to press. I imagine that the Americas won't be far behind.

Ten million EVs are forecasted to be on the road by 2025, which means that someone will end up with a healthy bottom line as a result of all this build-out. Right now, however, it's not an easy business. In fact, it's almost a surefire money *loser* in 2022, best suited for entrepreneurs who are *really* into EVs and don't mind living without a second hot tub.

Level 3 cheat sheet

While waiting to recharge your mental and vehicle batteries on that long road trip, try committing these Level 3 charging facts to memory:

Outlet: N/A. You don't install Level 3 chargers in your home, and even commercial installations require limitations such as permitting and land acquisition. Level 3 chargers have their own charging stations, owned and operated by a variety of companies.

Voltage/current: 400–800 volts DC

Connector type: SAE/CCS Combo, CHAdeMO, Tesla connector

Power: 75–350 kilowatts per hour

Rate of charge: 75–1000 miles of vehicle range per hour

Cost: Can vary widely. Most DC fast chargers bill somewhere between 32 and 36 cents per kilowatt-hour. In this case, adding 70 kilowatt-hours of power results in a fill-up cost of $22–$25. I have seen nonmember rates of 69 cents per kilowatt-hour.

THE YELLOW FUEL LIGHT AND THE PARADIGM SHIFT

As I'm sure you understand, a paradigm is a way to model your understanding of something, of making sense of your world. It a word whose Greek and Latin origins literally mean "to display side by side."

So then, let me compare the experience of fueling an ICE car and an EV *side by side.*

During your life as an ICE driver, your paradigm for fueling is that of a container filled with liquid. And, in fairness, that's exactly what a fuel tank is. You drive around until a yellow light displays on your dashboard, telling you that the liquid level in the container is getting low, and then find the closest (or least expensive) fueling station that sells the liquid that goes in the container, and then, in what is perhaps the most significant part of this fueling paradigm, you fill the container to 100 percent capacity. Most of the time, we simply fill 'er up.

You drive away and the yellow light disappears until the next time the container starts to run out of liquid.

Here's how that paradigm shifts when fueling an EV:

Usually, sitting in a cupholder, or affixed to the dash of the car that needs liquid added every so often, is a phone. Usually, that phone is plugged in to the car's charger so that the battery is ready to go. Does it matter if the phone is at 70 percent when you plug in the charger? 40 percent? 20 percent?

And what about taking it off the charger? Do you spend much time worrying about the charge level as long as you know it has plenty of power to take that selfie, play that podcast while working out, or post that tweet about this awesome book you happen to be reading about EVs? If the battery level is at 70 percent and you have a Zoom call coming up, do you reschedule until you've had time to charge the phone to 100 percent?

(continued)

(continued)

Of course not. Usually, you just plug in your phone, or unplug your phone, and don't give it a second thought.

This is all a somewhat lengthy — but, hopefully, valuable — way of saying that the paradigm of the EV is that of your phone. Under normal usage conditions, you don't use your phone until the red battery indicator shows up, and only then to think, "Hey, I should plug this thing in now, wait until it hits 100 percent capacity, and only then take it off the charger and get to phoning again."

Yet for some who are hesitant about EV adoption or practicality (and who say, "Ugh, they take for-e-e-ever to charge!"), the paradigm of gasoline fueling seems resistant to a shift.

But you're reading this book, so you're not that person. You know that, oh, about two to four days after taking delivery of your very first EV, the paradigm about fueling your car with electrons will indeed shift, and that within that first week, you'll very likely stop giving this whole charging business more than a passing thought.

Shortly after that first week of ownership, you'll drive by a roadside business that sells — what's this? — a flammable liquid. (Do people actually pour this flammable liquid into a tank and carry that liquid around with them constantly for the next several weeks?) People actually go to Costco and — am I seeing this right? — wait 20 or 30 minutes for the privilege of saving $3 or $4 on a container of liquid? Do these people not value their time?

What I'm saying is that you'll feel bewildered. You'll feel relief that it's not you out there in the freezing cold or blistering heat, inhaling toxic fumes. You'll think, "I'll never have to do that *ever* again."

When you have internalized this fact is when you'll know that that the paradigm has shifted for you, too.

» Calculating how much energy the battery can store

» Letting your phone handle the 100 percent charge

» Maximizing battery life/performance

» Replacing or recycling the battery (if needed)

» Seeing why swapping batteries isn't an option

» Knowing what to expect when replacing a battery pack

Chapter 8
Protecting the Battery (and Your Investment)

The battery pack is the most expensive part of an electric vehicle.

If you're feeling a sense of déjà vu, it's because I mention this fact elsewhere in this book. However, we *For Dummies* authors never assume that these books are consumed in the order specified in the table of contents. So, if this fact is new: Hi there, and welcome! If not, then great to see you again and thanks for reading this far.

Either way, now that you're here, let's talk about protecting that investment. And because the battery pack and vehicle range are inexorably linked, protecting the battery also means protecting the car's ability to move from point A to point B.

Cracking open this book — physically or otherwise — to this chapter means that you likely share common questions about battery life and battery range, as these are the two most common questions folks have when switching from ICE to electric.

Questions in this genre are understandable because EV batteries tend to be the same types of batteries — chemically speaking — as the ones either powering your phone or whatever other device you're using to read this section. You have no doubt noticed that your phone or laptop battery may not be able to power your device like it did when it was brand-new. You thus want to do everything you can to preserve both the range of the batteries and their lifespan.

So, if the preceding paragraph describes you, you've come to the right place.

Seeing Which Batteries Actually Power Cars

Strangely enough, I start this section with a look at what most people think of when they picture a car battery. (See Figure 8-1.)

FIGURE 8-1: A large toaster oven. No, wait — I guess this is a 12V car battery. It's been a minute since I've owned an ICE.

Sashkin / Adobe Stock

I likely don't have to tell you that virtually every ICE car contains a battery, and neither do I need to remind you of what happens if *that* battery dies. (I revisit the whole dead EV battery in a bit later in this chapter.)

A traditional ICE battery feels like a block of lead because, well, it mostly *is* lead — a lead-acid battery, to be exact, although the substance inside the casing is a mixture of lead and a water-based sulfuric acid electrolyte. The technology in these kinds of batteries dates back to the 1800s — this was the first rechargeable battery ever invented.

So why is the ICE battery still in use? Because lead-acid batteries are durable and long-lasting and, most significantly, they recharge themselves while the ICE engine is running. They provide 12-volt power to car electronics — like headlamps, sound system, and wipers, or just about anything that lights up, makes noise, or moves — operating these electronics even with the engine turned off. And, as many a stranded driver on a cold winter morning has found to their great dismay, the ICE battery also powers the vehicle starter motor. Without a working vehicle starter motor, it's difficult to, you know, *start* the car.

A 12-volt car battery typically weighs roughly 40 or 50 lbs., a weight I can still *comfortably* bicep curl with one arm — usually, right before heading to the beach. Get swole, feed the soul, bruh.

TECHNICAL STUFF

If the power at your house has ever gone out and your phone still held a cell signal, you likely have a lead-acid battery to thank for that. Large-format lead-acid batteries are frequently used as backup power supplies for cellphone towers.

Figure 8-2, on the other hand, shows a single EV battery.

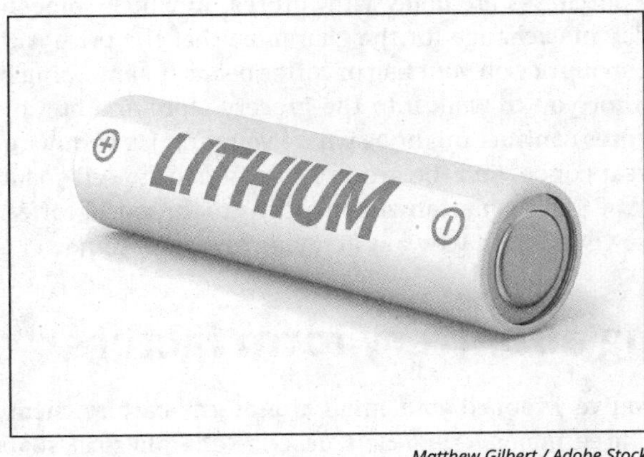

FIGURE 8-2: One EV battery cell. Or, if we're living in *The Matrix,* it's one human being providing energy to the computer overlords.

Matthew Gilbert / Adobe Stock

Humble-looking, isn't it? You probably have a few similar-looking batteries lying around in a junk drawer somewhere.

The lithium-ion (Li-ion) batteries are *also* rechargeable. One significant advantage lithium-ion provides is that these batteries are much lighter than their lead-acid counterparts. And please note that I'm talking like-for-like here rather than the entire weight of all batteries in an ICE when compared to an EV. What I'm saying is that when you compare a lithium-ion and a lead-acid battery of equal volume, the Li-ion will be lighter.

Perhaps the more significant reason Li-ion batteries are used to power electric cars is that they're more *efficient* than lead-acid.

Before we proceed, note that *Li-ion* can refer to a wide variety of battery *chemistries*. The most common Li-ion battery chemistry used in today's EVs is nickel manganese cobalt (NMC). However, some manufacturers (Rivian, Tesla, and others) are beginning to switch their lower-range battery pack chemistries to lithium iron phosphate (LFP) batteries because these battery chemistries require neither nickel nor cobalt — two metals that can be expensive or difficult to acquire. As far as this chapter is concerned, distinguishing between NMC and LFP is immaterial.

TIP

As I write this chapter, nickel has seen a significant price spike. As with the price of any commodity, the cost of raw materials affects final pricing. That said, worries about spot nickel prices causing massive EV price hikes is overblown. Without going into too much detail, that's (mostly) because a) EV manufacturers don't pay spot prices in the first place, and b) you can process low-quality nickel into high-quality nickel, which effectively sets a cap on the price of the high-quality nickel needed for NMC batteries.

REMEMBER

A *spot* price means the price to just go out and buy the commodity today. Many big commodity purchases are done with futures, in which someone pays for a fixed amount today in exchange for the guarantee that the price will remain the same in future. A spot price on your bag of coffee beans is something like $15 per pound. It costs $15 for you to walk into the grocery store and buy the pound of coffee today. A futures contract might be where you promise to buy 1,000 pounds of coffee over 5 years or so, and the grocery store agrees to sell you all 1,000 pound for a fixed price of $14, no matter what happens to the world coffee bean market over the next five years. (And why am I craving an espresso now?)

Looking at battery form factors

Now that you've wrapped your mind around the battery chemistry used by EVs, know that three major *form factors* describe the physical shape of these Li-ion batteries:

- **Prismatic:** Prismatic Li-ion cells consist of anodes, cathodes, and separators layered together like some kind of 7-layer bean dip that can store electricity. These layers are then pressed, folded, or rolled and then stuffed into metal or plastic casings. The benefit of a prismatic cell is that it can be custom-made, so it's ideal for applications like the laptop or phone batteries, where engineers have only a certain amount of space in which to work.

In EV applications, prismatic cells are most commonly used in hybrids, and for the same reason: There's only so much space to fit a battery into something that's ICE in all other respects.

- **Pouch:** To picture a pouch, think of a Capri Sun fruit punch drink or a squeezable yogurt. With a pouch Li-ion cell, cathodes and anodes are stacked and then wrapped in a flexible foil enclosure. This strategy reduces packaging weight and increases efficiency, which offers a particular benefit to EVs. Pouch cells are the preferred form factor used by cars like the Chevy Bolt, where pouches are used in a series and then surrounded by a metal casing.
- **Cylindrical:** You may hear the term *jelly roll* used to describe cylindrical cells, but when I think of what's inside a cylindrical battery, I end up thinking of a sushi roll. (Seriously, what is a jelly roll? Is it like a cinnamon roll?) Cylindrical cells sandwich together the sheets of anodes, separators, and cathodes, roll them, and pack them into — yep, a cylinder. As mentioned earlier in this chapter, you likely have a few examples of these items sitting in your junk drawer or your Xbox controller.

Looking at the advantages of cylindrical batteries

So, is one better than the other? To the end user of an EV, not really. Batteries are batteries. The full answer is more a matter of engineering preference, not of definitive fact.

That said, cylindrical cells *appear* to be winning the day. Yes, pouch cells are used by Chevy, Ford, and Hyundai, but I think it's telling that all three US-based, EV-only manufacturers (Tesla, Lucid, Rivian) have chosen cylindrical for their battery packs. These manufacturers have built their EVs from the ground up rather than retrofit existing car bodies (the Ford F-150, for example) with battery packs.

EV-only has gravitated toward cylindrical cells in their battery packs for what seems to be three key reasons:

- **Cost:** Because cylindrical cells are mass-produced by several competing manufacturers across the globe (like Samsung, LG Chem, CATL, and others — Tesla, like its CEO, rolls their own, as it were), the cost per kilowatt-hour is being driven down faster with cylindrical cells when compared to either prismawtic or pouch.

The industry goal for battery cost is $100 per kilowatt-hour. This is widely viewed as a tipping-point threshold, at which EVs achieve price parity with their ICE counterparts. The price is now close to $130 per kilowatt-hour, and when factoring in the price of gas and maintenance, I'd argue that we're at the point of price parity already. Nonetheless, competition and innovation continue to drive down the price of batteries per kilowatt-hour.

- **Structural rigidity:** Each cylindrical cell is encased in metal, passing along an engineering benefit to the overall vehicle chassis. Essentially, the battery structure becomes part of the car structure.
- **Performance:** When unleashing supercar-like 0-60 times down a drag strip or away from that stoplight so you can flex on that Camaro who pulled up next to you, ungodly amounts of power get unleashed very quickly, and battery packs with 5,000 to 7,000 cells have more points of contact for all that stored energy. The pouch cells used by a Chevy Bolt, on the other hand, might have 100 or so points of contact.

A fourth reason, cooling, would likely be argued by the chief engineering types who've opted for the cylindrical battery route. But there would be counterarguments from other chief engineering types who have chosen pouch cells, and I won't attempt to settle the debate in this space.

Although all cylindrical batteries share the same engineering genotype, not all present the same phenotype. Three main cylindrical specifications are now used in EV battery packs:

- **18650:** The 18650 name refers to the cell dimensions. An individual cell is 18 millimeters wide and 50 millimeters tall. If you were to see a picture of an unmarked 18650 with nothing next to it for reference, you couldn't be blamed for thinking you were looking at the same AA batteries used in your remote control.
- **2170:** The 2170 cells, which are 21 millimeters wide and 70 millimeters tall, began appearing in EVs starting with Tesla Model 3s and Model Ys. Lucid and Rivian both use 2170 cells, and Rivian's battery packs actually arrange them stacked atop one another.
- **4680:** The 4680 cells are — you're getting the hang of this now — 46 millimeters wide and 80 millimeters tall, significantly larger than either the 18650 or 2170 cells. (See Figure 8-3.) They were introduced to the EV world at Tesla's Battery Day in late 2020 and, what's more, they are theoretically cheaper to produce — it stands to reason that if you have a bigger cannister you can fit more electrode/cathode material in that cannister and use less raw materials on the cannister itself.

I'll come back to the 4680 in a bit, but for now, know that whether or not the 4680 is a significant step forward in terms of EV battery design or essentially a different sized wrapper around the same battery components is very much an open question.

However, it might not really matter. Another reported advantage is that the battery packs made from these cells become part of the vehicle frame, which in turn allows Tesla to use fewer reinforcement parts in creating the pack. Every little bit of cost savings helps.

Finally, Tesla has confirmed that the 4680 cells have been deployed in the Standard Range Model Ys coming off the line from the new Austin, TX Gigafactory.

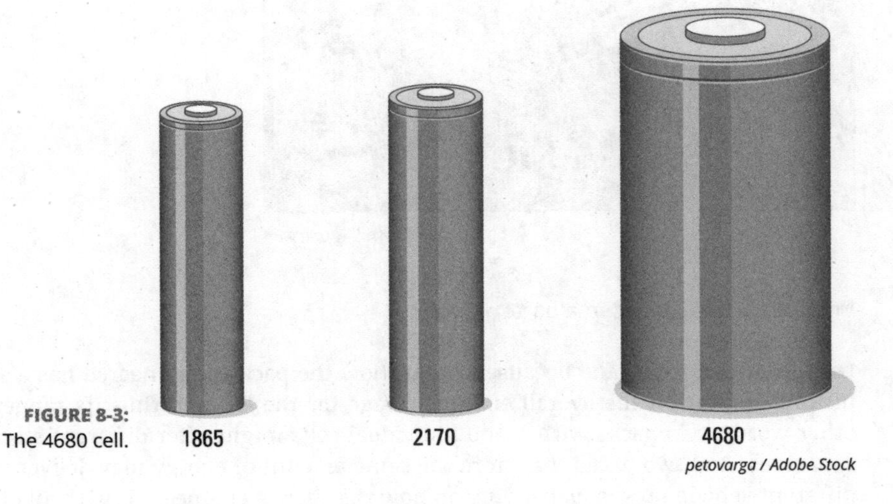

FIGURE 8-3: The 4680 cell.

petovarga / Adobe Stock

It takes a village

As you may have guessed by now (assuming you haven't skipped the preceding), a single battery cell cannot power the entire car. Heck, even most TV or gaming remote controls need multiple AA batteries, so getting to the grocery store or Grandma's house requires the combined power of several *thousand* battery cells all grouped together. (That's several thousand if we're talking about cylindrical cells, and between 50 and 100 if we're talking about Li-ion pouches.)

So why not cover the car roof in solar panels and use the sun to charge the battery pack? There are lots of reasons why this is impractical. One has to do with manufacturing the roof and then stapling it onto the car frame without damaging the cells. The other is weight — as in, it would add a lot of weight to an already heavy car. The *other* other is that the surface area of a car just won't allow more than a tiny trickle of the sun's energy to make it to the battery. Solar calls in/on a vehicle could theoretically power the car electronics, but certainly wouldn't provide enough juice to charge the car.

This arrangement of battery cells is known as a *battery pack*, as shown in Figure 8-4 with some of the structural casing cut away.

FIGURE 8-4: One EV battery pack.

Picture Alliance / Getty Images

What all is happening in a battery pack?

Lots, as it turns out, and because of that, how the pack is engineered has a significant impact on the overall energy storage for the EV, and thus its range. In other words, two packs with 4,000 individual cells might offer different storage capacities. And two packs that store the same amount of energy may deliver very different vehicle ranges, depending on how the pack is engineered (with all other factors equal).

In any event, the role that battery packs play in EV performance is hard to overstate. For starters, it's the piece of technology that's responsible for battery cooling, for safety in the case of a thermal event (defined in the chapter on EV safety), and for charge management.

Often, battery packs become part of the car's *structure*, impacting the safety of vehicle occupants in a crash, as I discuss in Chapter 3. (The short version is that the batteries absorb some of the mechanical energy of an impact.)

Battery packs are so influential on overall EV function, in fact, that they're mentioned prominently in branding and awareness campaigns. For instance, when General Motors talks about the Ultium platform, it's referring specifically to the modular (pouch-based) battery system. Likewise, Telsa has applied for a patent for its new structural battery pack, built around the new 4680 cells.

Battery pack capacity

No doubt about it: Capacity of the battery pack does have a significant impact on several aspects of vehicle performance.

As I mention earlier in this chapter, the 2170 cell is bigger than the 18650, and to greatly simplify, more energy can be stored in the bigger cell. Specifically, it's about 50 percent larger in terms of volume, but can deliver almost double the current. The higher current of the 2170 also allows for higher charging speeds when compared with the 18650 — a 2019 Model 3s can charge faster at a Supercharger than a 2019 Model X, which uses 18650s.

Because the cells are larger and store more energy, fewer cells are needed for the same-size battery pack. Think of the sushi roll I mention earlier, in the section "Seeing Which Batteries Actually Power Cars." With the 2170 "roll," you need less seaweed, less rice, and so on, so you save on the costs of these raw materials.

Again, let me acknowledge that what I've just stated is an oversimplification, but there's a reason both Rivian and Lucid have built their battery packs around this larger 2170 cell configuration when the 18650 is just as commercially available.

Now, on to the 4680.

The 4680 cell is larger still than the 2170. As such, you'd expect it to store more energy, sure, but Tesla claims that the 4680 will store up to five *times* more energy than a single 18650 cell, which will result in a 16 percent increase in range for two battery packs of equal size (in kilowatt-hours; more about that later in this chapter).

The more significant factor, however, is cost. Tesla says the 4680 packs will cost 50 percent less to build.

Whether the cost savings are delivered by the 4680 cells or something else, I cannot overstate how critical these cost savings are to being able to sell larger vehicles, like pickup trucks (the Rivian R1T, the Ford F-150 Lightning, and the Tesla Cybertruck, to be exact), much less semitrucks (from Tesla, Nikola, Volvo, or someone else) for even a meager profit.

Given the current state of electric vehicle engineering and the price of batteries, it simply takes too many batteries to move the amount of mass in a pickup truck (never mind a semi) without having the cost of those batteries launch the vehicle sticker price well into the six figures. Reasonably priced EV pickups (this is all relative, of course) are simply a math problem that can't be solved without the purported advantages of the 4680.

I'll deal with future battery tech in Chapter 10, so the problems may also be solved in a myriad of other ways.

In any event, on the other end of the BEV spectrum is the much-anticipated and long promised "$25,000 Tesla."

The Tesla community has repeatedly referred to this future Tesla as the Model 2, complete with fan-fiction renderings of what a Model 2 might look like, although Elon Musk has just as repeatedly refuted that name. No matter what the vehicle ends up being called, a $25,000 Tesla is *entirely* dependent on Tesla producing a less expensive battery pack, because the company seems to have little interest in producing a smaller, Nissan Leaf-ish car that might have only 80–125 miles of range.

All that said, though, it's a matter of debate whether the 4680 is truly a breakthrough in battery cell technology — or whether it's a case of more sizzle than steak. Is it a better battery, or simply better packaging?

As of now, there's no clear-cut answer, mostly because we haven't seen that many 4680 cars on the road yet, which in turn means we don't have enough data about a) performance or b) cost. We'll find out that information only by digging through quarterly earnings reports that are still several years away.

One thing we can very much understand today is the overall size of the battery pack. The unit of measurement — kilowatt-hour — has already been mentioned in this section, so let's take a moment to examine and understand the carrying capacity of an EV battery pack.

Counting Up the Kilowatt-Hours

Given what I hear almost daily, vehicle range is the principal concern of almost everyone considering an EV. People want to know how far the car travels on a full charge.

But what exactly a full charge means can vary widely from car to car, and even from model to model within a product line.

It has a lot to do with the size of the battery pack.

It helps to understand that battery pack capacity is expressed as a function of the kilowatt-hours (kWh) stored at full charge. What's a kilowatt-hour? It's the ability to deliver one kilowatt of electricity for one hour. Your electricity is sold to you in units of kilowatt-hours, with the price per kilowatt-hour ranging anywhere between 12 cents and 35 cents in the US, depending on location and time of use.

For example, a 1,000-watt microwave oven can run for one hour on one kilowatt. Okay, fine, but you're not popping corn for a marathon of every Marvel movie since *Iron Man* — you just want to know how far an EV will go on a single kilowatt-hour of energy. That number can vary (and greatly, as an expression of percentage), but the general ballpark is 3 to 4.5 miles per kilowatt-hour. I know — that isn't far, but if you look at the numbers in the previous paragraph, you'll conclude that most people will be able to travel 1 mile for about 3 or 4 cents.

That's pretty cheap transportation, if you ask me, and I guess that by reading this book, you sorta are.

In any event, that's why EVs today need battery packs that can store between 75 and 125 kilowatt-hours of energy. A bit of quick math on a 75 kWh battery pack says that, at 4 miles per kilowatt-hour, a full battery will get you 300 miles, and I hope that's the last time I'll use three numbers in a single sentence.

The Tesla Model 3 Long Range AWD, for instance, achieves a higher-than-average 4.4 miles per kilowatt-hour, meaning that its 82 kWh battery pack yields up to 358 miles of range.

REMEMBER

As with ICE gas mileage, how you drive plays a significant role in determining vehicle range. If you're in the habit of ripping away from every stoplight, each stop a new opportunity to experience the EV smile that comes with flooring the accelerator, well then, your mileage may indeed vary.

So it's simple, then: A larger battery pack is better, yes?

Well . . . as you'll see in a moment, the answer to that question is the same as it is for most of life's questions: "It depends."

As a general rule, a battery with a higher kilowatt-hour number is analogous to an ICE vehicle with a larger gas tank — in theory (and, for the most part, in practice), you'll be able to drive the higher kilowatt-hour EV further between "fill-ups" — even though filling up is the wrong paradigm. (For more on paradigm shifts, see Chapter 7.)

But the weight and the aerodynamics of the car also have a huge impact, which is a topic we'll get into next.

Yet another EV rabbit hole that I invite you to journey down at your leisure features a discussion of EV *efficiency*. The car's efficiency is a function of several inputs, such as motor efficiency, energy management, tire size, and aerodynamics.

Again, I'm not going to unpack the nuances of vehicle efficiency here. As a general rule for now, Tesla, Hyundai, and Lucid score well in the kilowatt-hour-per-mile department, whereas Audi, Porsche (Volkswagen, in other words), Jaguar, and Toyota are further down the list.

Kilowatt-hours in vehicle badging — that's so 2018

And while we're on the subject of battery pack size, Tesla used to feature its kilowatt-hour ratings front and center on its vehicle models.

When purchasing a used 2017 Model X, you can choose between the 75D and the 90D. The 75D comes with a 75 kWh battery pack and approximately 237 miles of range, whereas the 90D has a 90 kWh pack and 257 miles of range. The P and D, incidentally, refer to the Performance variant and Dual motor configuration (which means all-wheel drive, or AWD).

Now, all manufacturers, Tesla included, have stopped mentioning the kilowatt-hour capacity in the vehicle model name. A used 2020 Model X is now either Performance (and 272 miles of range) or Long Range (and 328 miles of range). Both use a 100 kWh battery pack.

The Chevy Bolt uses a 65 kWh pack, Ford's Mach-E offers either a 68 or 88 kWh pack option across four variants, and other models, like the Ioniq or Polestar, use 78 kWh packs. In fact, 75–85 kWh now seems to be pretty much the sweet spot for many of the sedan/hatchback models now on the market.

Meanwhile, an even larger 135 kWh battery pack is used in the 2022 Rivian R1T. Yet the range of this larger battery pack allows for an EPA-estimated range of "only" 316 miles. For comparison, Tesla's Model Y has a 75 kWh battery pack, and EPA ranges of 244, 303, and 330 miles, depending on the configuration you choose (rear-wheel drive/RWD, Long Range all-wheel drive/AWD, or Performance).

What's the difference between all these cars?

Well, look at them. One is a low-to-the-ground, swept-back luxury sedan, one is a midsize SUV, and one is a pickup truck. Without even discussing the fact that all use slightly different motors, it just takes more energy to push a boxy pickup truck through the air than it does a sedan with the profile of an aircraft wing.

It can also be important to keep in mind that batteries take up space. Take the Lucid Air Dream Edition sedan, for example. The Air Dream uses a 118 kWh battery pack, which can propel the extremely aerodynamic sedan down the road for nearly 600 miles. The more base-model Lucid Air Pure has an 88 kWh battery, which is still good for 406 miles. However, the larger battery pack of the Dream Edition actually means a bit less cabin space for passengers. The floor of the Pure is lower than in the Dream, a situation that might be noticed by taller back-seat passengers. (The 93 kWh pack of the Air Touring variant also has a lower floor and thus more legroom, much to the delight of one owner, as shown in Figure 8-5.)

FIGURE 8-5: More legroom = joy.

Getting down to the nitty-gritty

Time to start bottom-lining this whole battery storage discussion.

Because it can take a little digging to find out a car's battery pack size, and because the number of kilowatt-hours stored in the battery pack won't provide that much information in terms of range anyway, a more helpful number to search for when assessing EVs is the EPA range, which is a neutral assessment of how far the vehicle travels on a full charge.

The EPA's methodology tests both operating range and energy consumption in laboratory conditions, which means they're placed on a dynamometer — think of a gym treadmill, but for a car (and also much harder to pronounce than *treadmill*) — and then run through several simulations of both city and highway driving. An ICE vehicle would begin with a full tank of gas and is run to empty; an EV starts the test with a full battery and runs until the battery taps out.

The current range champion is Lucid, which, yes, happens to be my employer, but as I explore throughout this section, not all range is created equal — and as I'll tell anybody who asks, most drivers need only about 20 to 30 miles of daily range, anyway. For my money, I don't think it's worth agonizing about EPA range stats after you get into the 250–300 mile ballpark.

TIP

A helpful website to bookmark in your browser's EV Shopping folder is `https://fueleconomy.gov`. There you can look up range specs for current and past EV models. You can also get this information from the Monroney sticker on the vehicle, although modern EV shopping can entirely bypass the whole "looking at the sticker price" part of car shopping, in favor of online configuration and purchase. Nonetheless, the law requires every new car to display this sticker on a side window. (And, yes, it really is called a Monroney sticker, named after Senator Mike Monroney, from the great state of Oklahoma, who sponsored the Automobile Information Disclosure Act of 1958, which mandated its use.)

On a final note, the EPA also issues EVs a miles-per-gallon "equivalent" rating, which you can see on its website and on window stickers as the MPGe. The MPGe is what the EPA says is the EV's *effective* miles per gallon of gas rating. In other words, whereas MPG estimates how far a sedan will travel on one gallon of gas, the MPGe estimates how far an EV will travel on the *equivalent* amount of energy contained in a gallon of gas. As it happens, the energy equivalent of one gallon of gas is 33.7 kilowatt-hours of electricity. So, from a cost perspective, if you factor in about 4 miles per kilowatt-hour as an average, EVs will cost the same on a per/mile basis of a car getting between 100 and 130 MPG. One "gallon" of electricity will take you about 120 miles. Not bad.

And with that, I've brought you full circle with my discussion of filling up a battery. Now let's talk about why doing so is generally a bad idea.

Recognizing That Too Much of a Good Thing Is a Bad Thing

Though the basic operation of a laptop and EV battery are identical, they differ significantly in how they're charged and discharged. Let's look at a few of these differences in more detail:

- **Car batteries don't really charge to 100.** The worst thing you can do for the longevity of a Li-ion battery is to cycle it from 0 to 100 percent state of charge.

So car battery packs don't really allow for this. An EV battery pack has a built-in buffer, which means that even if the display reads 100 percent, you haven't quite charged the pack quite that high. Conversely, even if you're stranded on the side of the road with a battery reading of 0, they have a small bit of juice still remaining (although that doesn't mean the car will operate).

Manufacturers engineer longevity into their battery packs by adding this buffer of not quite 0 percent when drained and not quite 100 percent when "fully" charged.

- **Car batteries need to last longer than phones do.** I can almost hear the thought running through your head: "So, what you're saying, Brian, is that phone manufacturers *could* extend the lifespan of their phones if they practiced the same battery management techniques as car manufacturers. Is that right?"

I can't answer this question from an engineering standpoint (nor will I bother to research it — I have enough on my plate, with the car stuff), and I certainly can't answer it from an *intent* standpoint. What I can state is that from a *consumer behavior* standpoint, many people upgrade their phones every two or three years. If, by the end of year 3 a laptop battery lasts 50 percent as long as it did when new, then I guess that's just how the turns table. (Hat tip to Michael Scott, of *The Office*.) Neither manufacturer nor consumer seems to pay it much mind.

An EV battery simply can't behave like your average phone battery does. Consumers would definitely pay attention to a battery swap of $20,000 to $40,000 every three years.

The good news, however, is that no fewer than three purely electric cars have been around for roughly ten years now, from which we can gain a good barometer of EV battery longevity. The first two are the Tesla Model S and the Model X. Tesla gathers a large quantity of data about its vehicles, and, according to the company's public impact reports, owners can expect between 10 to 15 percent battery degradation between 150,000 and 200,000 miles.

In other words, a Tesla battery pack with a range of 300 miles when new will have a range of 255 to 270 miles after about ten years of typical driving. Additionally, this self-reported data is consistent with crowd-sourced reports from Tesla owners. (For all the details, check out `www.tesla.com/ns_videos/2020-tesla-impact-report.pdf`.)

The third vehicle for which we have a long track record of battery information is the Nissan Leaf. I recall being introduced to the Leaf during the 2013 Tour de France, and the reason I bring this up is to highlight the fact that the Leaf has been driven all across Europe long before the Tesla arrived, and their battery longevity experiences have been nearly identical. According to surveys, more than 90 percent of UK Leaf owners have experienced no problems with their vehicles.

Dealing with Battery Death

Here's another question that frequently comes up with EV consider-ers: What happens if I run out of battery?

Answer: The same thing that happens when you run out of gas.

How the empty tank and empty battery differ, at least for the next few years or so until charging infrastructure improves, is that if your ICE car dies, you can have someone bring you a gallon or two in a gas can. One of the properties that makes gas such a miracle energy source is how portable it is.

If your battery dies in an EV, however, you have to get towed to the nearest charger. That said, remember that EVs are computers on wheels. Every single EV I've ever laid eyes on does a fantastic job of presenting battery telemetry to the driver. You should know exactly where you stand at all times — how much further you can go on your current charge, how many miles you're getting per kilowatt hour, and much more. Many modern EVs will also provide directions to the nearest charger.

CARRYING AROUND A BATTERY "GAS CAN" IN YOUR TRUNK

A few companies offer portable charging solutions. If you're willing to lug around a very heavy box in the back of your car, you can get between 20 and 40 miles, should you find yourself in a pickle.

One such backup battery vendor is ZipCharge, and you can order yours at `www.zipcharge.global`. Another is the Roadie, by EvUnited (`https://evunited.com`).

I imagine that these products will see the most use by towing or other roadside assistance services. As with a gas can, someone providing assistance can bring you a few miles of juice to get you to the nearest charger.

It's not for me, because portable backup batteries are both heavy and expensive, but on the other hand, I haven't given range a second thought since about my second week with an EV. We all have different levels of comfort, and if this helps, then you do you.

As a bonus, the backup battery in your trunk won't smell up your car.

Once again, this situation kind of goes back to the whole paradigm shift associated with "fueling" an electric. EV drivers just don't drive around until they see a yellow low-fuel light. For daily driving, they start every day with a full "tank" and just don't give much thought to charging up.

Going the Extra Mile

What are the best tactics to preserve battery pack health, to give it the best possible odds of providing range for longer than the car's sheet metal holds together?

Here are a few of the best tips, as collected from owner's manuals from across the spectrum of EV manufacturers. You will notice several common themes.

- **Keep it plugged in:** According to Tesla, the most important factor (to) "preserve the high-voltage battery is to LEAVE YOUR VEHICLE PLUGGED IN when you are not using it." The battery works best when charged regularly.

  And yes, Tesla uses ALL CAPS, the written equivalent of shouting.

  Rivian concurs with this: "Leave the vehicle plugged in if you don't plan on driving for a long time. When parked, the vehicle uses some power to maintain battery health."

- **Avoid overcharging it:** Another universal recommendation is not to charge to 100 percent unless absolutely necessary. (It never is.) I made this point in a separate section of this chapter, but here it is, straight from Hyundai:

  "Electric cars have . . . a battery management system that avoids them being charged and discharged at the extreme state of charge. Even though a full charge will give you the maximum operating time, it is never a good idea for the overall lifespan of your battery."

- **Avoid exclusive use of Level 3 charging:** It used to be conventional wisdom — and common advice — that Level 2 charging works best for everyday use, and best for the life of the battery. Why? Heat. Higher voltages mean higher heat while charging, and heat can stress the battery pack.

  There's a significant "however" with this bit of advice which is this: a recent study by the Idaho National Laboratory found that an EV's battery pack deteriorates faster if its *only* source of charging is DC fast charging. This is rarely the case with real world EV ownership. What's more, the study found that the battery degradation when *only* using Level 3 charging was relatively minor. (For more on this, see: `www.greencars.com/post/why-dc-fast-charging-reduces-ev-battery-life`.)

- **When possible, park in the shade:** Generally, it's not a great idea to park an EV uncovered for long periods in very hot conditions. Lucid's manual says the following:

  "Extreme temperatures can damage the battery pack. If possible, avoid parking in direct sunlight, especially on hot, sunny days. Lucid also recommends that you keep your vehicle sheltered or parked in a garage whenever possible in extremely cold weather."

- **Let the battery cool before charging:** Again, it's that heat issue. Avoid it, if at all possible. If you can park in the shade on a hot summer afternoon, do so.

Seeing Where EV Batteries Go When They Die

A natural question that arises from EV purchase considerations is this: What happens to all those batteries after the car has reached the end of its life? Since a lot of 'em look like AA batteries, do they get thrown in the trash when they're dead, like my remote batteries do? (By the way, don't do this! You can recycle used batteries at most office supply stores (thank you, Staples and Office Depot) or at hardware stores (and thank you, Lowes and Home Depot).

Alternatively, what happens to all those batteries even when the battery pack isn't dead? (As you've seen in this chapter though, EV battery packs don't really die the same way that a phone battery might.) What happens, for instance, when a Lucid Air is involved in a crash that insurance writes off as a total loss, yet most of the batteries are unaffected? In this case, several thousand batteries are in perfectly good shape. What happens to those?

Or, what about the batteries on a Chevy Bolt that's been recalled? Let's say the battery pack is completely swapped out as part of the recall. What happens to the replaced batteries — or, more specifically, what happens to all the *raw material* in those batteries?

To answer these questions, let me make a few introductions.

Dirk Spiers and Spiers new technology

Dirk is Dutch by birth and Oklahoman by choice, although why someone from Holland would choose to live in Oklahoma remains one of life's enduring mysteries. Maybe he lost a bet, or maybe he's in the Dutch witness protection program.

While we're on the topic of origin stories, the scoop on Dirk is that he once pressed a General Motors head of development about his plans for Chevy Volt batteries after they died. (The GM head of development probably thought Dirk was pressing, that is. Spiers, being Dutch, probably thought he was simply being direct.) In any event, the answer Spiers received at the time was, "We don't really have a plan," and thus a business was born.

He now runs the *very* directly named Spiers New Technologies, which tackles what Spiers describes as the four Rs of battery services: repair, remanufacturing, refurbishing, and repurposing. (If you take a look at the Spiers New Technology website, you'll find Rs practically falling out of the virtual sky: Add root cause analysis and recycling preparation to the full menu of available services.)

For more on Spiers New Technologies, check out `www.spiersnewtechnologies.com`.

JB Straubel and Redwood Materials

JB gets to put "cofounder of Tesla Motors" on his resume. It's only one more multibillion-dollar EV carmaker than *I've* founded, mind you, but I guess that's not nothing.

The short version of Straubel's resume goes like this: While serving as Tesla's CTO, JB built up the battery engineering team. After leaving Tesla, JB founded a company called Redwood Materials, which takes used Li-ion batteries of all manner — phones, laptops, even power tools — and "un-manufactures" them into raw material. As Straubel points out, perhaps the largest lithium and cobalt mines in the Western hemisphere are spread throughout the junk drawers of America.

In short, Redwood's overarching aim is to help create a circular battery-supply chain, saying that it has "an appreciation for the incredible scale of materials involved in the transition of the whole world to sustainable energy, and we know it is not too early to plan for "un-manufacturing" the many gigawatt-hours of batteries being built today."

And these are just two of the many men and women who are representative of two companies entering this space. Others are getting into the battery recycling game as well, companies like these (at the time of this writing):

- **Hydro (Norway):** `www.hydro.com`
- **Li-Cycle (Canada):** `https://li-cycle.com`
- **Northvolt (Sweden):** `https://northvolt.com`

According to the Institute of Electric and Electronics Engineers (IEEE), several dozen start-ups in North America are getting into EV battery recycling for fun and profit.

TECHNICAL STUFF

Speaking of recycling: Li-Cycle is now repurposing an old Eastman Kodak building complex in Rochester, New York, turning it into a $175 million battery recycling plant. It will be the largest Li-ion recycling plant in North America.

So the good news is that smart people have rolled up their sleeves and are hard at work on the issue of EV battery recycling.

The even better news is that recycling your EV battery pack 8 to 12 years from now will be business-to-business (B2B) in nature, so you shouldn't have to mess with it, just like you don't have to mess with what happens to your car after you trade it in at the dealership. Sometime in the 2030s, Ford will happily take your 10-year old Mustang Mach E as a trade, and if it ends up salvaging the vehicle or swapping out the batteries, chances are very high that someone will show up with a satchel full of cash so that they can get their hands on all that sweet, sweet rare earth metal.

For more on Redwood Materials, check out `www.redwoodmaterials.com`.

As a final note on EV battery recycling and reuse, consider that another potential use of an EV battery is for grid storage. A degraded EV battery might not be able to produce the range or torque that a vehicle driver demands, but it appears that there will be a lively market for used EV batteries for grid storage.

This is one of many direct reuse scenarios, where batteries collected from EVs taken off the road (such as with an old car, or an insurance write off) would be re-deployed.

For example, as far back as 2015, BMW began piloting a program called iCharge-Forward, whereby *in-vehicle* BMW batteries were put to use as flexible grid storage for Pacific Gas and Electric. You can read more about it here: `www.pgecurrents.com/2017/06/08/pge-bmw-pilot-successfully-demonstrates-electric-vehicles-as-an-effective-grid-resource/`.

More recent projects have included a 1.9 megawatt-hour (MWh) installation of Audi e-tron batteries in Berlin, and a 1 MWh grid storage solution in the UK consisting of old Nissan Leaf batteries. (Source: `https://blog.ucsusa.org/hanjiro-ambrose/the-second-life-of-used-ev-batteries/`.)

In the state of California, it's estimated that almost 50,000 EVs will be retired from roadways by 2027. Using a very conservative estimate of 25 kWh per battery pack, there would be over 1 GWh/year of available storage. That's a lot of solar power that can be soaked in by the EVs on Cali's road's today. Talk about a closed loop system. (Now, if there were just a way to make it rain out here.)

All told, these second-life battery applications are still finding their footing, but do present clear cost advantages over acquiring new batteries at grid scale.

Your EV recycling to-do list

What's on your EV recycling to-do list? I mean, what do you have to keep in mind starting the day of your maiden voyage?

Nothing!

Just drive your EV. In 2033 or thereabouts, remember that Brian Culp told you about Redwood Materials and a Dutchman who has the misfortune, either by fate or by choice, of having to spend summers in Oklahoma. Maybe those two businesses will still be around, or maybe not.

But one thing for sure is that the metal in your vehicle's batteries will still be around, and chances are very, *very* good that someone will take those old batteries and harvest the raw material within, if for no other reason than it pays to do so. And, by the 2030s, it's also very likely that economies of scale will have driven down the cost of recycling batteries even further.

What's more, when the cells in a battery pack have reached the end of useful life for the EV, they might still be serviceable for applications such as home energy storage. This notion hasn't gotten much further than the drawing board just yet, but researchers at MIT have suggested that depleted EV batteries might find a second life as an affordable storage solution for utilities generating electricity from clean sources like solar or wind. For more on that topic, check out the page at this link: `https://spectrum.ieee.org/lithiumion-battery-recycling-finally-takes-off-in-north-america-and-europe`.

Oh, and I'll close this section with one more reason for optimism when it comes to recycling EV batteries: Unlike the AA batteries in your remote control, EV battery packs are extraordinarily hard to chuck into the garbage bin and then leave out by the curb for Wednesday pickup. (Incidentally, don't throw away AA batteries, people!)

IT'S NOT AN XBOX CONTROLLER

Manufacturers once toyed with the idea of a quick-change battery pack, with the idea that you could pull into something resembling one of those express oil-change locations, steer into a stall, hoist your EV on a lift, drop the spent battery pack, and attach a new one — and off you go, 15 minutes from empty to full.

The issue, however, is one of practicality and scale.

In terms of practicality: Why bother changing a battery pack that can last 300,000 miles or more? Why would a new-car owner spend their money on a new battery pack — again, the most expensive part of the EV — only to place that battery pack in someone else's car a few weeks later?

In terms of scale: It would be very, very, very difficult in terms of the capital and operational expense placed on a company to maintain a network of battery-swap locations. In fact, the stations would come in most handy in places along Interstate 80, like Sidney, Nebraska. I've sat at the Sidney, Nebraska, Tesla supercharger for 30 minutes during a road trip. Would it have been nice to pull into a Jiffy Swap and be done in 15 minutes? Sure. Did the extra 15 minutes make any measurable difference during a 10-hour day of driving? It did not.

The bottom line is that a quick-swap battery pack is a classic example of a solution looking for a problem. These are cars, not remote controls.

Even assuming that you know how to extract the 1,000 lb. battery pack from your EV — something that should be attempted only by trained technicians because EV packs can easily crush or electrocute — and then load it into the back of yet another vehicle for transport to the local landfill, be advised that your local landfill won't even accept used battery packs. They're too big of a fire hazard.

So I'll leave you on this thought: It's not entirely clear what to do today in terms of recycling an EV battery. When the time comes, I'm confident that the path forward will be clear. There will be too many EVs on the road by then, and the financial incentive to recycle all those batteries will be too great for things to be otherwise.

All that said, there is one instance where it does make sense to consider your battery replacement options in the more immediate term. The next section deals with that situation.

Dealing with Battery Pack Replacements

First, let me reiterate: You should never need to do this for a pure battery electric car. Really!

A BEV battery pack should last the lifetime of the car. Seriously, if you're reading this chapter before making your first EV purchase, take a snap of this page, save it to your Favorites or create a calendar event for five years or so in the future, and come back to it then.

With that task out of the way, the process for replacing a severely degraded battery pack would start with a call to your dealer or (in the case of Rivian, Lucid, or Tesla) to the manufacturer's customer support team. And, if you're making this call less than eight years into ownership, the battery pack is likely covered under the warranty. (Have I mentioned that you don't need to read this section yet?)

If you're in a situation not covered under warranty for some reason — perhaps you've purchased a salvaged car or have been using the vehicle for drag races that you post to YouTube — it's simply a matter of telling the dealer or customer service rep that you need a new battery pack and asking about the cost.

But again, if you're buying an electric vehicle today and you drive it extensively for the next ten years, putting 300,000 miles or so on the chassis, and then you decide that you want to hang on to the car for sentimental reasons, just call and ask.

Who knows? By that time, battery technology may have advanced to the point where the cost of battery replacement is a restively minor cost. (Have I mentioned you truly don't need to read . . .— oh, never mind.)

When replacing a battery makes sense: The case for hybrids

The most common instance of replacing a battery pack for the next couple of years will be in the example of a hybrid vehicle — think of replacing a Prius pouch-style battery. In this instance, you can likely find a used or refurbished battery from a third party and have them do the installation. If this is you, know that a used or refurbished battery will cost about half the price of a new one, and will probably make a lot more sense because the car you're buying it for will also have a few years under its belt.

So let's say that you want to squeeze a few more years out of a 2015 Toyota Prius whose battery won't get you nearly as far as it used to.

For somewhere in the ballpark of $1,400, you can have a new battery installed in that car, using one of the options listed in the next section. Better yet, many of these refurbished batteries come with a generous warranty (some even offer a lifetime guarantee), and installation techs are mobile — they do all work while you're in the kitchen making a frittata.

Reviewing battery replacement options

Here are a few options to consider when shopping for a hybrid battery replacement:

- **Greentec Auto,** www.greentecauto.com**:** Billing itself as "the fastest-growing hybrid battery replacement company," Greentec Auto has 18 locations across the US, covering over 85 percent of the population. Greentec offers mobile installations and an unlimited mileage warranty.
- **Hybrid 2Go,** https://hybrid2go.com**:** Hybrid 2Go operates in California and Arizona, offering mobile installation and a lifetime warranty on its refurbished batteries. It restores the function of your hybrid's battery by thoroughly testing capacity and then rebuilding using good cells where needed.
- **Bumblebee Batteries,** https://bumblebeebatteries.com**:** Bumblebee offers similar battery swaps as the other firms mentioned here, except that it uses a "BeeMax 16-bit extreme-precision cell matching process," whatever that means. I guess it's a good thing the company has a FAQ page, for clearing it up. It's apparently an algorithm that helps optimize cells when refurbishing, by sampling voltage, amperage, impedance, charge and discharge rates, capacity, and other indicators of battery health.

  Anyway, you get a 3-year warranty, and the company will ship hybrid battery packs directly to your door, provide DIY battery swap videos, and connect you with a service center if you're not the DIY sort.

In sum, opting for a refurbished battery for your older hybrid vehicle is a savvy move that can extend the life of your car and the breathing room in your bank account. A listing of the firms offering this service in *your* area should be a quick Internet search away.

The fallback, of course, is to contact a dealership that services your hybrid. Battery replacement services are offered at almost any Honda, Toyota, Chevy, Ford, or Hyundai dealership.

» Finding out where to get your EV serviced

» Caring for and protecting the exterior and interior

» Breathing a (cleaner) sigh of relief with some carefree maintenance

Chapter 9
Keeping Your Car in Top Shape

The EV revolution means automobiles with fewer moving parts, which in turn means fewer points of failure, which in turn should mean you'll spend less of your time and money keeping the EV on the road.

Of course, "less service" doesn't mean "no service."

EVs are still machines made of glass and steel, held together with rivets and glue, and nothing lasts forever. Additionally, these machines require a significant upfront investment (or CapEx, to my infrastructure planners in the audience), and so it certainly makes good financial sense to spend a little on routine maintenance rather than a lot on major repairs (that's called lowering OpEx in the infrastructure planning biz).

Speaking of major: As you're probably aware, there are significant differences between an ICE automobile and an EV. Keeping on top of the maintenance schedule is always recommended no matter what kind of vehicle gets you from Point A to Point B. It's just that the components you'll need to maintain for each type of car are like night and day. As a result, the place you go for that maintenance may also vary (and, in some instances, that place might even be your garage).

This chapter, then, mostly serves as a maintenance primer for new EV owners. It covers the items that commonly appear on upkeep checklists and covers the options for getting those items properly serviced.

Getting Your Hands Dirty (or Not)

For the most part, a car is a car and almost any automotive service center can handle a good chunk of your EV service.

The biggest difference between an EV and an ICE vehicle isn't so much where you get it serviced, but rather *what* you get serviced. As has been mentioned elsewhere in this book, you won't ever pull into a Jiffy Lube — or at least not for a 15-minute oil change. You'll have neither spark plugs serviced nor timing belts replaced. You won't swap out fuel filters or stop for an emission check or fail to remember to check the antifreeze levels in the winter.

Not doing things never sounded like this much fun, yeah?

Most of the routine maintenance on an EV has to do with the items that connect the drivetrain to the road and provide power to that drivetrain. They are:

- Tires/wheels
- Brakes
- Battery

Let's look at each in the sections that follow.

Dealing with tires

We'll start with the most significant item of the three components connecting car to the ground.

Maybe it's because I spent much of my life playing sports, but I have a special affinity for shoes — of any variety, really, but especially of the sports variety. When I was a younger man, I agonized over the most comfortable (and yes, best-looking) shoes for basketball, soccer, and, of course, baseball. As a man now of a certain age, I still sweat the details when choosing a pair for either work or play. Why bother? Because all the activities associated with sport start with feet on the ground, and so whatever is between your foot and the ground is perhaps the most important piece of equipment you'll choose.

MY SHOE DEFENSE

I should perhaps mention that a) I'm not one of those sneakerheads who has a YouTube channel and more pairs of shoes than there are days in a month, and b) I do nonetheless have more pairs of shoes than any one person needs. Also, I acknowledge here that shoes are not environmentally friendly. That's why one of my favorite pairs is the Ultraboost Made to Be Remade, a shoe that's not only extraordinarily comfortable but whose lifecycle is also circular. That is, the entire shoe can be sent back to Adidas to be recycled: upper, midsole — the entire product is ground up and remanufactured. It's a process I applaud, much more so than Nike's Space Hippie line, which takes waste material that was going to be thrown away, shapes that material into a shoe, sells it to you, and then lets you throw it away instead.

So it is with tires. They're the shoes on your car. They're the bond between a multi-ton hunk of metal and the earth. Without tires (and gravity, in all fairness), your car would just fly off into outer space and no one outside of Elon himself wants cars in low earth orbit.

And just like shoes — especially of the sports variety — tires should be checked regularly for signs of wear. Like shoes, they need to be replaced whenever those signs of wear point to a potential loss of function.

As a rule, an EV's added weight means that it's a bit harder on tires than are internal combustion vehicles. (This isn't the case with something like a Prius or Nissan Leaf. It's all about the curb weight — the weight of the vehicle including a full tank of fuel (which doesn't apply) and all standard equipment.) A Tesla Model 3 weighs about 4,000 lbs., and a Prius Prime weighs 3,370. A Toyota Corolla weighs 2,850 lbs.

So, when feeding and caring for your EV, make sure to set a calendar reminder to check those tires (or, if you're setting this reminder from Manchester, in the UK, the home of my second-favorite soccer team, check the tyres). When that reminder notification goes off, you'll be on the lookout for cracks and uneven wear, and you'll also want to keep an eye on the tread depth indicators. See Figure 9-1 if you're unsure what I'm talking about.

What are tread depth indicators? They're these little cross marks spaced evenly along the main grooves in the tread. When new, a tire has a tread depth of 10/32 or 11/32 inches (8 to 9 millimeters). When worn, the tire tread wears down to 2/32 inches (1.6 millimeters), at which time the tread is even with the tread depth indicators. When that happens, it's time to replace the tires. Tires with 1.6 mm of tread or less are considered unsafe for driving, because the tread wear translates into a loss of grip, which in turn impacts braking distance and overall vehicle control. In wet conditions, your vehicle runs an increased risk of hydroplaning when the tires are worn.

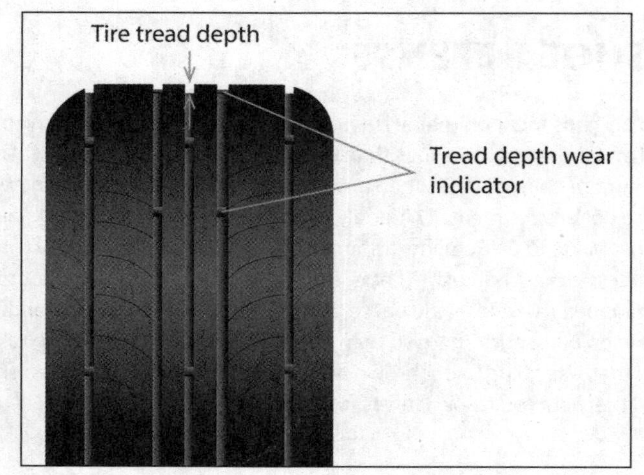

FIGURE 9-1: Check the tread depth indicators to know when it's time to change tires.

Pankaj_Digari / Adobe Stock

Reading tire markings

Tires have a wealth of information printed right on the sidewall — it's the law. But you could stare at a sidewall all day and not be able to figure out what all the numbers and abbreviations mean — unless you have some kind of fancy decoder ring provided by an automobile or tire manufacturer. (See Figure 9-2.)

FIGURE 9-2: Got your decoder ring?

Zigmunds / Adobe Stock

Well, guess what?

Here's the information to look for in the tire sidewall:

- **Tire category:** P indicates that the tire is intended for passenger vehicles. HL indicates that the tire is a high-load-rated tire (something that may not be shown on certain tires).
- **Tire width**: This 3-digit number gives the width of the tire in millimeters, from sidewall edge to sidewall edge. So, if a tire is marked P245/35R21, the tire width is 245 mm.
- **Aspect ratio:** This 2-digit number, also known as the *profile,* gives the sidewall height as a percentage of the tire width. If the tire width is 245 mm and the aspect ratio is 35, the sidewall height is 85.75mm.
- **Tire construction:** R indicates that the tire is of *r*adial ply construction. So, if a tire is marked P245/35R21, the *R* represents the *r*adial ply construction part.
- **Wheel diameter:** This 2-digit number is the diameter of the wheel rim in inches. So, if a tire is marked P245/35R21, the wheel diameter is 21 inches.
- **Load index:** This 2- or 3-digit number is the tire's *load index,* a measurement of how much weight each tire can support. This number is not always shown.
- **Speed rating:** The speed rating, when stated, denotes the maximum speed at which the tire should be used for extended periods. The ratings range from 99 mph (160 km/h) to 186 mph (300 km/h).
- **US DOT *t*ire *i*dentification *n*umber (TIN):** Regulations require that the TIN begin with the letters *DOT* and be followed by two numbers or letters that indicate where it was manufactured. The last four numbers represent the week and year the tire was built. For example, the number 1706 means the 17th week of 2006. The other numbers are marketing codes used at the manufacturer's discretion. This information can be used to contact consumers if a tire defect requires a recall.
- **Maximum permissible inflation pressure:** Maximum permissible inflation pressure means the maximum cold inflation pressure to which a tire may be inflated. It's okay for the tire pressure to exceed this value when it's warmed up, but it must be at or below this pressure when the tire is cold.

  In the US, tire pressure is measured in Pounds per Square Inch, or **PSI**. In places that use the metric system, it's measures in kilopascals, or **kPa**.
- **Treadwear grade:** This is number required by the government that indicates the tire's expected wear rate. A grade of 300 means that the tire will wear three times as long as a tire graded at 100. However, these grades are assigned by the tire manufacturers, not independent third parties.

The treadwear grade is part of the US Department of Transportation's tire rating system that was established in the 1970s to help buyers better understand passenger vehicle tires. (As you can see in this list, there's a lot to keep track of.) The Uniform Tire Quality Grade (UTQG) incudes treadwear, temperature, and traction ratings.

- **Traction grade:** This letter indicates a tire's ability to stop on wet pavement.
- **Temperature grade:** This letter indicates a tire's heat resistance grading.
- **Tire composition and materials:** The number of plies in both the tread area and the sidewall area indicates how many layers of rubber-coated material (plies, in other words) make up the structure of the tire. Information is also provided about the type of materials used.
- **Maximum tire load:** This is the maximum load that can be carried by the tire.
- **International tire approval marks:** If applicable. Before products can be sold in many countries, manufacturers are required to test and certify they meet government standards, and any certification logos may also appear. Figure 9-3 shows an example of the China Compulsory Certification (CCC) mark.

FIGURE 9-3: No, it doesn't stand for the Civilian Conservation Corps.

Whew! I *told* you it was a wealth of information. An embarrassment of riches, informationally speaking — and maybe stiff knees — await those willing to squat down and have a look.

For most drivers, **the most important thing to look for is the maximum inflation pressure,** which again is the maximum amount the tire should be inflated to **when cold.**

Notice that I used a boldface font in that previous sentence, and if you know anything about me as an author, dear reader, it's that I don't ever yell at my audience in ALL CAPS and that I am sparing when it comes to the use of bold text. So the bold is there for a reason.

Correct tire pressure is important for all cars, but is particularly true for EVs. To gain the most efficient use of the energy coming out of the battery, you want the tires at their optimal **cold** PSI. (There's that bold text again.)

That said, sometimes it can be difficult to locate or see the PSI numbers on the tire sidewall. Maybe you've parked in the shade and left your reading glasses on the kitchen counter and the tire's PSI text just happens to be upside down, about an inch from the pavement, and you're just not feeling like doing a yoga tripod pose next to the convenience store's air pump.

The good news is that you might never have to look at the tire sidewall to find this information. Every manufacturer also has to put a tire-loading information sticker on the inside of the driver's-side door (along its B-pillar). This label also contains the maximum tire pressure information, and it can usually be read much more readily than the black raised text against the black backdrop of the tire itself. Figure 9-4 shows where it's located on every car. Again, it's the law.

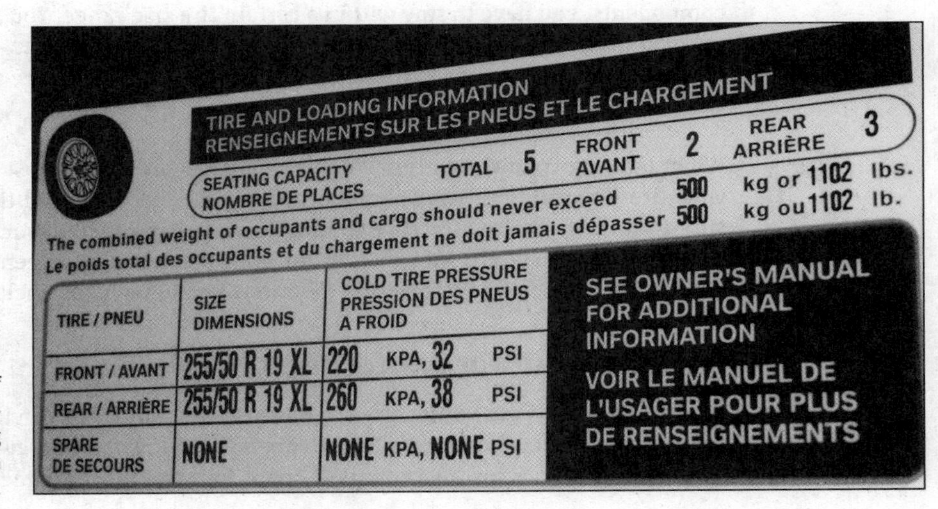

FIGURE 9-4: This label can save you lots of battery range — and from being forced to do yoga in a parking lot.

The sticker contains the following information:

- Maximum vehicle capacity weight in kilograms (kg) and pounds (lbs.)
- Maximum number of occupant seating positions in the vehicle
- The size of the tires originally fitted to the vehicle
- The *cold* inflation pressures for the original specification of front and rear tires

Well, what if you swap out the tires? Doesn't that invalidate all that sticker information?

In theory, it can, but it's a very rare instance where that's the case. Almost always, you'll change out tires like-for-like. You might change from the Michelins that came supplied with the vehicle to a product from Continental or Hankook, for example, and the tread pattern will likewise change, but the tire size (and, most often, speed rating) will remain the same. Unless you modify other drivetrain components, you have to stay within a certain tire size range. You might swap out 19-inch tires for 200-inch tires, but you won't swap them for 22-inch tires. As a result, the PSI recommendations between different tire sizes generally stays the same.

When lighting conditions are favorable, it's probably a good idea to take a pic of your tire sidewall with your phone if you have indeed swapped the originals. Be sure to favorite the pic for easy finding later on, and Bob's your uncle the next time you need max PSI information for your tyres, which is generally how you spell *tires* if used in a sentence that mentions Bob being your uncle.

Swapping out for winter tires

EV owners should use winter tires to increase traction when driving in sustained temperatures below 50 degrees F (10 degrees C), or in snowy or icy conditions. For winter tires, always install a complete set of four tires at the same time.

All winter tires should be the same diameter, brand, and construction and have the same tread pattern on all four wheels.

Choosing an EV-specific tire

All the exhilarating torque from an electric motor has a bit of a downside: physics.

Torque that's both powerful and instant, while also torquing below a body of significant mass — namely, the battery pack — causes stress on tires over and above that of a gasoline vehicle. In layman's terms, habitually treating your EV like a drag racer can chew up tires as though . . . well, as though you were racing your car on a track all day.

Fortunately, certain tires are specifically designed to meet the additional stresses placed on them by EVs. Generally speaking, these tires offer more durable rubber compounds, minimal rolling resistance, and even reduced noise transference, all of which help EV drivers enjoy a longer range, more energy-efficient ride.

The following list gives you a sampling of EV-specific tire options that will give you a feel for what to look for when swapping out your E's first set of tires:

- **Continental EcoContact 6:** Continental's EV-specific line ". . . addresses the specific issues raised by the increased weight of EVs and hybrids. [The tire] helps to balance the need for low rolling resistance, high grip, and sturdiness."
- **Michelin Energy Saver A/S:** Michelin says of its Energy Saver that it's the "ultimate fuel-efficient passenger car tire . . . ideally suited for drivers looking to maximize fuel economy without compromising long tread life and all-season safety."
- **Toyo Versado Eco:** The Versado Eco from Toyo is "an all-season, low rolling resistance luxury performance tire designed for hybrids and other environmentally friendly vehicles. Using a combination of innovative technology and materials friendly to the environment, the Versado Eco delivers low rolling resistance without sacrificing tread life, luxury, or performance."

Putting on the brakes

Over in Chapter 2, I look at how EVs use regenerative braking to slow themselves down and add energy back into the battery pack.

Using this feature, most drivers will drive from place to place using one-foot driving — careful drivers will be able to conduct their daily commutes essentially without ever using the mechanical brakes. It's a very cool technology; I've made countless trips around town and without a single tap of brakes.

The bottom line is that you may go many years without ever replacing the brakes and/or brake pads.

To flesh out this EV brake maintenance with a few numbers, I'll use Teslas as an example, because we have a wealth of data about the upkeep for these cars. On its Car Maintenance website (`www.tesla.com/support/car-maintenance`), Tesla suggests a brake fluid check every two years, with replacement as needed.

Meanwhile, according to Tesla owner forums (and websites that report information collected via these forums), a typical owner experience is that the Tesla's brakes will last between 100,000 and 150,000 miles, which will get most owners through year 7.

So, if you're buying a new EV immediately after buying this book, go ahead and set yourself a calendar reminder for 2030 or so. I just indulged in a rather lengthy discussion on tires, but thanks to an electric vehicle's ability to stop itself during most routine driving without having to touch the brakes, I can keep this section brief.

For a little further reading about regenerative braking and how it saves on brake maintenance, I leave you this: `www.torquenews.com/15475/why-you-would-seldom-need-brake-pad-replacement-tesla-models-wonders-regenerative-braking`.

Dealing with batteries

This section looks only at the topic of battery maintenance. And because there are no moving parts, batteries should need very little in the way of year-to-year attention.

As far as *replacing* the battery — or, more accurately, the remote possibility of ever *needing* to replace the battery — that's a subject I cover extensively in Chapter 8.

Although EV battery packs don't need much maintenance in the traditional sense of the word — a mechanic puts your car on a lift and takes something out or off, services it, and then puts it back again — most EV batteries can be calibrated from time to time.

A common method involves these steps:

1. Drive the car until you're in the single-digit percentage range.
2. Let it sleep for about an hour by turning off all standby functions, like Tesla's Sentry mode. Disconnect your phone app and walk away.
3. Charge the battery back to 100 percent and then drive the vehicle as soon as you're able.

All of what I've just described is one method of resetting the EV's *battery management system (BMS)*, which is simply the software used to manage battery consumption.

Do you ever *need* to do this? Probably not. I include it as something that's worth looking into if you find the car's reported battery range isn't meeting expectations — if you're a year into ownership of your 300-mile range EV, and the car at full charge is reporting only 240 miles of range. (A loss of 20 miles of range after one year of ownership would not be cause for alarm.)

All EVs use BMS software, so the trick is to figure out how to reset the BMS for your car. Sometimes the procedure can be found easily enough in an online search; sometimes the knowledge is passed along, more word-of-mouth style, in owner forums or obtained from a service center or dealership advisor.

Beyond that, most of what falls under the category of battery maintenance is up to you, not some random mechanic.

Optimizing battery health

These are the three most significant suggestions for optimizing the health of the battery:

- **Maintain a state of charge above 10 percent.** If you can help it, try to keep the battery's state of charge (SOC) above 10 percent. Once in a while is no big deal (you may want to drive to under 10 percent for a BMS calibration), but try to make it the exception, not the rule.

  I'll discuss the reason why in the next bullet point.

- **Avoid regularly charging to 100 percent.** The reason why is that lithium-ion batteries are stressed when sitting either at a full charge or an empty charge. (It's not good for your phone or laptop battery, either, but it's less noticeable because the devices are typically replaced far more often than a car. That said, can I get an "amen" if your laptop is more than three years old and you've encountered a severely degraded battery.)

  The easy way to avoid charging to 100 percent is to set the car's charging limit using the app you were asked to install at delivery. A general recommendation to which almost all manufacturers pass along is to charge the vehicle to 80 percent for daily driving.

  80 percent SOC should be more than plenty given the range of today's EVs. A typical commuting day for most in the US will use 10 percent-15 percent of the battery for an EV with a 300-mile range.

You can also set charging limits using the in-car infotainment system. Every manufacturer will include instructions about how to do this. (Here. For example, are Ford's directions about how to set the charging limit on a Mach-E: `www.ford.com/support/how-tos/electric-vehicles/ev-range/how-do-i-set-the-maximum-charge-level-for-my-electric-vehicle/`.)

» **Protect the battery from high temperatures.** This advice includes avoiding DC fast charging (DCFC) whenever possible, because DCFC sessions put a lot of heat into the battery as a by-product of the high-voltage current.

Interestingly, lithium batteries perform their best in roughly the same temperature ranges as human beings. If it's too hot or cold, most EV battery management software systems will either cool or warm the pack.

However, heating and cooling requires energy, which is pulled from the pack itself. If the pack is spending energy heating or cooling while parked, this means less range for your next drive.

Mind you, this is not a battery *longevity* issue, but if you want to avoid this behavior, park in the shade on a hot day and in a climate-controlled garage in colder weather.

The bottom line is to keep the battery's daily state of charge set to about 80 percent to help ensure that your EV battery never needs maintenance.

Tracking Down EV Service Locations

Though the battery might not need maintenance, the other two items in this section certainly will — rarely for the brakes and more frequently for the tires. But more components surround your typical EV driver than just these three items, such as cabin air filters, heating and cooling systems, power seating, windshields, windows, side mirrors, and — not least of all — the electronic systems that control this computer on wheels.

But EVs are new. The many old-timey garages along the famous Route 66 in the US may be terribly Instagrammable (see Figure 9-5) but might not be a terribly great place to service your EV. What are the options, then, for EV owners who are curious about *where* to take their cars?

FIGURE 9-5: A great place for pics of your road trip and an oil change, but unlikely to be a great place for EV service.

travelview / Adobe Stock

Finding service at a dealership

At least in the near term — say, any time before 2025 — if your EV comes from one of the legacy automakers (Volkswagen, Ford, Hyundai, GM), it's a safe bet that you'll be headed back to the same dealership whenever the car needs service. And once that EV is there, it's also a safe bet that the cost of the dealer having to have additional tools and expertise in order to service that electric drivetrain will be passed along to you. That's just the way it will go for EV drivers when taking their cars into places that exist primarily to service gasoline vehicles.

And guess what? The EV-only manufacturers still charge for their mobile services (more about that in a minute). I don't have much else to say about dealer EV service at this point, because we're still early in the world's uptake of EVs and most of the ones out there are too early in their lifecycle to need much in the way of service.

REMEMBER

There seems to be a notion that EVs will significantly cut into a dealership's service revenue ("Dealers will never sell an oil change again!"), but I think that sentiment is more than a bit overblown. First of all, dealers don't make money on $35 oil changes that cost them $200 in materials and overhead. Oil changes tend to be loss leaders for other types of service business. (Think back to your own experiences with oil changes at a dealership and see whether you can think of a time when an additional service wasn't recommended.) Second, EVs will still have their windshields cracked, get into fender benders, or need their tires rotated. Dealers won't do these things for free.

Going mobile (service, that is)

With the elimination of dealerships from some EV-only manufacturers comes the elimination of taking your car to a dealership for service.

Enter the mobile vehicle care service offered by the likes of Rivian, Lucid, and Tesla.

Using the mobile service from one of these automakers to take care of routine tasks — like air filter replacements, cracked windows, or punctured tires — is as simple as ordering a pizza. For the most part, all you do is open your car's app and click on the Service menu or icon and then you're off and running. (See Figure 9-6.) Tell the app what service you need, and most of the time you can get scheduled right then and there.

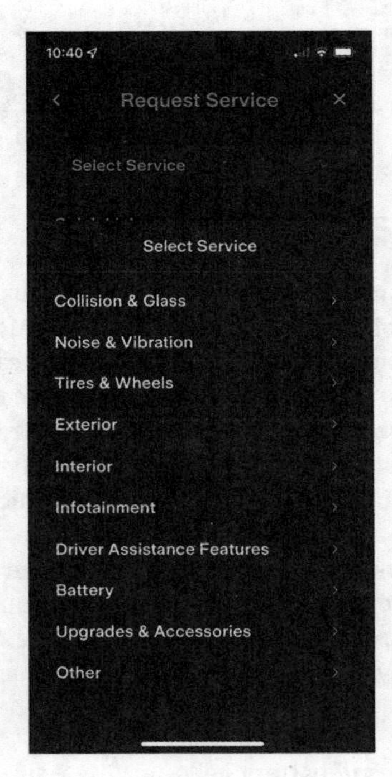

FIGURE 9-6: There's an app for that.

For example, while on a weekend getaway with my family, our Model Y took a rock thrown by a passing truck that cracked the passenger-side window. While still on our trip, we scheduled a mobile appointment for the following Thursday. While we worked from home that day, the technician arrived and worked on our Y while it was still parked in our garage, and we paid using the credit card linked to our Tesla account. To say that it was easier for us than making a trip to the gas station is to understate how much time and trouble it takes to fill up a tank of gas. We've had more difficulty completing an accurate takeout order from DoorDash.

Now, the fact that we live in the same city where Tesla has a giant factory may have something to do with the level of service we receive. I've heard plenty of anecdotes about service experiences that have been what people in a boring corporate meeting would call *suboptimal.* Yet the fact remains that, when done right, mobile auto service is how *all* auto services should be performed.

And it's not just Tesla that makes house calls. Lucid and Rivian also have fleets of mobile service technicians ready and able to perform a majority of services while your car is either at home or at work.

Rivian even goes one step further, saying on its website that "if more extensive care is required, we'll pick up your vehicle, bring it to a Rivian service location and return it to you, all with a flexible transportation program." Lucid also offers vehicle transport services if the problem can't be resolved in a customer's driveway.

Oh, and Lucid vans arrive with not only parts and tools but also coffee. (I hope it's Peet's Major Dickason's blend, which is my coffee of choice in the Lucid breakroom — dark blend, full-bodied, earthy. Chef's kiss to the Dickason's blend, which you can order at www.peets.com, by the way.) Figure 9-7 shows a Lucid mobile service van in all its glory.

FIGURE 9-7: Is this a mobile service van or a coffee food truck that also happens to be able to rotate tires? I guess it depends on which side of the transaction you're on.

Lucid Motors

Benefiting from third-party services

Dealerships might still be hit-or-miss when it comes to providing EV service. Fortunately, the only feature on EVs that requires routine service is the tires, and those can be rotated and/or changed out at just about any third-party business that sells tires — Pep Boys, Tires Plus, any Tire Rack third-party provider, and even your local Costco. Just skip the nitrogen, please. (Are we really back at Costco again? Really?)

Beyond that, I don't have much to add about third-party service at this point. What's available from third parties depends a great deal on your location and the popularity of EVs in that part of the world. If you're in California or Norway, you can likely find several shops in your area with EV mechanics at the ready. If you're in South Dakota or Uruguay, your third-party options are likely considerably more limited.

Doing it yourself with DIY service (right-to-repair)

One option for those needing to service their vehicles is to go the DIY route. Here again is where the law is on your side, thanks to the concept known as right-to-repair, or R2R.

Right-to-repair is kind of a fascinating topic.

Entire chapters, and even entire books, could be dedicated to describing the history and implications of right-to-repair, but it essentially refers to a consumer's ability — their right — to fix consumer products they've bought instead of being left with two rather unpalatable alternatives: Either get the thing serviced *only* by the manufacturer, which has specialized tools, parts, and expertise (diagnostic software, in many cases) or just throw the item into a landfill. Think phones, computers, farm equipment, and, yes, automobiles.

This concept isn't specific to autos, of course, but it does provide some background about right-to-repair: A rather notorious point of contention recently has involved Sonos speakers. A software "upgrade" made legacy items (devices made in years prior, in other words) unable to be repurposed without getting Sonos' permission first.

For more reading to help provide context about right-to-repair in general, check out: `www.vice.com/en/article/3a8dpn/sonos-makes-it-clear-you-no-longer-own-the-things-you-buy`.

WARNING

I hope this concept is self-evident for any would-be EV mechanic, but the voltages stored in a battery pack are lethal, not to mention that "dropping" the pack can kill you if you're standing underneath a vehicle on a car lift and decide to remove the battery. The technicians who work on electric vehicles take specialized training first to understand electricity and how to safely work in and around EV battery packs, and then they take additional training about how to safely remove battery packs when necessary. Neither the voltages nor the weight of the battery packs can be taken lightly, pun very much not intended.

In any event, auto service and right-to-repair are closely interlinked, even though the subject of right-to-repair is broader than just autos. Massachusetts has been the first US state to pass a R2R law, and at the time of this writing, legislation is working its way through both the US Congress and the California state legislature. (Because California is the largest EV market in the US, whatever it adopts will no doubt become a de facto standard.)

Beyond that, it won't take you much searching on either Reddit or YouTube to find accounts of expensive electric vehicle repairs (mostly of the Tesla variety because they're generally the ones old enough to need out-of-warranty repair work) that can be done much less expensively by third parties with the right tools and skill.

The bottom line is that there's a reason third-party auto service centers exist. There's no way that dealerships could service all of the United States' 300 million cars on the road, and in a free market, consumers deserve options about where and how to repair their machinery.

TIP

If you want to go a little further down the R2R rabbit hole, Marques Brownlee has a video overview I can recommend, at `https://youtu.be/RTbrXiIzUt4`.

As electric cars evolve into just "cars," the status quo is almost certain to change. Watch this space.

Now, although you might not have the knowhow or the equipment to rotate the tires on your EV, EVs have some care-and-protection tasks that individual owners can handle quite admirably by themselves. I discuss some of these tasks in the next section.

Keeping Your EV Exterior in the Pink

It's certainly no secret, and it certainly gets a mention several times in this book, that EVs aren't exactly free. As a reminder, though, along with the price tag comes the possibility of driving a very low-maintenance car for many years. It's not unreasonable to expect an EV's drivetrain to outlast its body.

So, how best to protect that EV body, and thus how best to protect the upfront investment?

Opting for paint protection film

Many EV buyers opt for an outer layer of protection called *paint protection film (PPF)*, which offers exterior protection from scratches that are part of life in a parking lot — the errant coat zipper or jean rivet or the stray shopping cart, for example — and is also useful in combating paint chips that routinely occur as a result of everyday driving. During almost every highway trip, your car is pelted by rocks and other small road debris thrown from the tires of other vehicles, not to mention the innumerable bug splats and/or bird droppings that are part of everyday driving (and parking).

If you closely examine the front of your current car, chances are you'll see dozens of small scratches and chips in the paint caused by the small collisions at high speed.

Paint protection film — sometimes referred to as Clear Bra, Clear Wrap, or Invisible Shield — is a thin layer of clear polyurethane that protects your car from this kind of damage. (See Figure 9-8.)

FIGURE 9-8: Kind of like lamination . . . for the entire car.

kamonrat / Adobe Stock

Because it does that, it keeps your car looking better longer. Buying PPF for your EV can be a rather expensive proposition, with costs for the best materials in the $4,000 ballpark. But Porsche Taycans, Lucid Airs, and Model S's all can run well into the six figures, so getting a quality PPF done by a professional (I would recommend DIY dental work before trying PPF by yourself) will likely be less than you pay in sales tax.

As for who's offering PPF, the big names here are 3M (with its Scotchgard brand), Xpel, and Suntek.

TIP

If you go the PPF route, it's probably worth investing the extra money to get either a self-healing PPF or an instant-healing PPF. (Self-healing paint protection films have multiple layers that allow for repairing surface-level damage when they are heated, which in most cases just involves being exposed to direct sunlight. At a micro level, the film becomes "molten" and fills in the scratch in the film.)

Entry-level PPFs are susceptible to fading and yellowing as they are exposed to UV rays. Newer PPFs have coatings that block UV light rather than absorb it, keeping them looking better for much longer. A good PPF is warrantied for 7 to 10 years.

Wrapping up the exterior

Another intriguing exterior protection option that can also radically change the very color scheme of your car is to apply a vehicle wrap. With a wrap, you can not only protect the paint but also change the color from factory white to very much non-factory matte black, cornflower blue, or even, um . . . rainbow chrome, I guess? If you're viewing the image in black and white, take my word for it: rainbow chrome is a very accurate description. (See Figure 9-9.)

Vinyl wrapping is not free, of course — wrapping the typical EV will run you about $2,000 to $3,000 — but one of the coolest properties of a car wrap is that it allows you to change your mind relatively easily. The PPF is to car color what henna is to tattoos.

The wrapping procedure is done by essentially applying large vinyl stickers to your car's surface. The trick — and it is indeed tricky — is to do the application without introducing bubbles, creases, or other imperfections.

In addition to changing the car's color, the vinyl wrap protects the underlying paintwork from minor stone chips and paint fading. And, when you decide to either change the color or revert to the original paint, you can easily remove the wrap.

The typical lifespan for a vehicle wrap is around 5-7 years, assuming proper care and maintenance.

FIGURE 9-9: Don't drive naked.

Source: Signature Custom Wraps, `https://www.signaturecustomwraps.com` / last accessed July 11, 2022

As with PPF, you *could* do a car wrap yourself by just buying the raw materials plus all the tools needed for the job. As with PPF, I strongly advise you otherwise. You can also make your own bread, ferment your own wine, and grow your own vegetable garden — and, yes, I've done all three of those things — but ultimately, I trust professionals who have better equipment and expertise than me when I want a really good sandwich that doesn't take me a day to make and costs three times as much.

There are many, many small businesses that specialized in applying PPF and vinyl wraps. In terms of the vinyl wrap manufacturers, the playing field is significantly narrower.

Your local shop will likely recommend a product from one of the following firms:

- **Avery Dennison:** Yes, this is the company that makes envelope stickers.
- **3M:** And yes, this is the company that makes Scotch tape.
- **Hexis Car Vinyl:** Makes the Skintac line of vinyl wraps.
- **Arlon:** SLX and Fusion wraps lead their product lineup.

Coating in ceramic

Ceramic coating is a silica-based liquid polymer that's applied either directly to the paint surface or even over the top of an existing paint protection film. After a few days, it cures into a hard protective shell that can be effective for several years.

I've invested in ceramic coating for both my family's Model 3 and Model Y and have found it to be very much worth the cost of application, which was about $1,000 each time.

Ceramic coatings are extremely hydrophobic, meaning they repel water — water on a ceramic-coated car slides right off, and this same characteristic of repelling water also prevents water stains, road grime, bird droppings, and other substances from "sticking" to the paint job.

Ceramic-coated cars are easy to maintain — they can be washed with water and a towel. No soap necessary. Additionally, most places that apply ceramic coating will advise you not to take your car through anything other than a touchless car wash (no spinning brushes, in other words).

Oh, and if you're not in the market for a brand-new car that runs 50-60k, and I don't blame you if you aren't, then a ceramic coating is probably still a good investment for a $20,000 car that you want to keep looking nice with a minimum of time, effort, and the expense of car washes, waxes, and so on. Yes, $1,000 will buy a lot of trips to the car wash, so it's not like you'll be making money. But the time and trouble saved is certainly worth considering.

Keeping Your EV Interior Shipshape

There's nothing radically different about the interior of an EV when compared to other vehicles — in fact, there's nothing different about them at all. Like their ICE counterparts, an EV interior is made up of glass, carpet, upholstery, sometimes leather, and sometimes vegan leather, which is leather made only from organic material, unlike synthetic leather, which is made from petroleum-based polyvinyl chloride or polyurethane.

Synthetic leather is not good for the environment, at either the time of manufacture or the time of disposal. Vegan leather can be made from items like mushrooms, cactus, algae, pineapple leaves, bark cloth, or even paper. It's cruelty-free and animal friendly, and its manufacture is associated with fewer CO^2 emissions than animal leather. And while we're here, anything associated with cows is not good for the environment. We all know this, yeah? Yeah.

Caring for your seat

Caring for your seat is a product of what the seat material is made from. Cloth seats should be kept free of loose dirt and other debris. (*Other* in this context almost certainly means old French fries — you know who you are.) Seats can be cleaned with virtually any upholstery cleaner. With leather seats, it's mostly the same story, except for the cleaner you use. The trick is to keep the seats out of the sun, if possible, and also prevent them from drying out. For vegan leather, you usually only need a soft wet cloth, warm water, and nondetergent soap, if necessary.

As with any car seat, cleaning up a spill as soon as possible helps make it easier to remove stains.

TIP

I have it on good authority (and the authority is my wife, who works at Tesla and hears the scuttlebutt passed down from seating engineers) that it's a good idea to keep a pack of baby wipes handy in your glovebox. Apparently, baby wipes are particularly effective at cleaning up the occasional spill on their particular vegan seat composition. Oh, and I guess the wipes can be useful with, you know, babies and stuff.

Cleaning that big screen

That big touch screen is both the control center and main information display for your electric vehicle, so keeping it free of dust and fingerprints can not only make it easier to see but also may aid in the functionality, or at least the ease with which you change car functions (switching stations on the radio or turning on seat heaters, for example).

But what do you clean these things with? Or, more specifically, is there anything you *shouldn't* clean them with? Like, are you going to mess up the ability to turn off the windshield wipers or end up turning on your seat heaters on a 90-degree day? More crucially, should you use ordinary window cleaner on the touch screen, or might that result in a Tesla fart mode "malfunction" during a first date?

The simple answer to the previous question is no — car touchscreens aren't that fragile.

That said, there's never a need to reach for an ammonia-based household window-and-mirror cleaner when simply ridding the touch screen from fingerprints. Manufacturers will always advise against it.

Essentially, treat the touchscreen the way you would treat a pair of eyeglasses. Or, failing that, treat the screen the way you would treat a Kindle or an iPad. That is, when cleaning, turn the thing off or put it into sleep mode (see the owner's manual for instructions), and then gently wipe the screen with a microfiber cloth — or use a combination of a mild cleaning spray and a microfiber cloth. And by *mild*, I mean mild. A car's touch screen can be cleaned with simple distilled water for most fingerprints and dust removal jobs. (I recommend not using tap water, because it can leave behind residue from dissolved minerals.)

Otherwise, as mentioned, reach for ammonia- or alcohol-free cleaning solutions.

You can even make your own cleaning spray by using one part distilled water and one part white vinegar. The vinegar's acidity dissolves substances like oils and sugars to make wiping the screen easier than using water alone. If you're doing this, which my wife and I have done for everyday household cleaning (yes, we're *those* people), I highly recommend adding a few drops of essential oil to cut the vinegar odor. Lemon, tea tree, and lavender are particularly helpful options.

Organizing your trunk/frunk

With electrics comes extra storage space, thanks in part to the fact that the car doesn't need room to carry around an internal combustion engine. The extra storage can be under the hood or under the trunk (in an area called the subtrunk) or, in the case of a Rivian RT1, between the truck bed and the cabin, an area Rivian calls the *gear tunnel storage compartment.*

Figure 9-10 shows a frunk from a Ford F-150 Lightning. And you thought you're supposed to put stuff in the pickup bed. Meanwhile, almost all EVs contain a subtrunk under the regular trunk for extra storage, and manufacturers like Rivian have introduced new spaces like the gear tunnel storage compartment. (See Figure 9-11.)

What all these little — and not so little — nooks and crannies of storage have in common is that there's almost always a product available to help you subdivide that space and organize life essentials within — stuff like mobile charging kits, gym bags, golf clubs, luggage, first aid kits, and groceries.

It's difficult to tell you exactly which products and companies will offer which storage and organizing solutions by the time you read this, but I can tell you with nearly 100 percent certainty that searching for the keywords *car make and model* and *storage compartment* will yield multiple options.

FIGURE 9-10: Gym bags, groceries, surfboards, secret *Death Star* plans — really, what *doesn't* fit in the many storage compartments offered by an EV? (Part I)

FIGURE 9-11. Gym bags, groceries, surfboards, secret *Death Star* plans — really, what *doesn't* fit in the many storage compartments offered by an EV? (Part II)

Roschetzky Photography / Shutterstock.com

To use just one random example, Figure 9-12 shown one of the results from a search for the term *Tesla Model 3 rear trunk organizer.* Items like these will bring joy to your inner Marie Kondo.

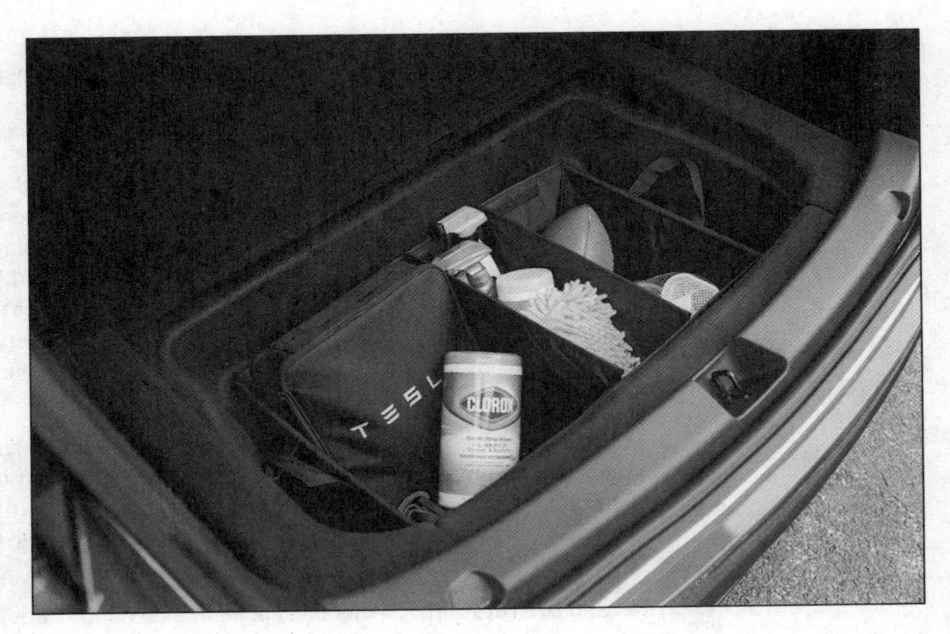

FIGURE 9-12: Google is your friend.

Knowing What You Don't Need to Do

Recap time.

First of all, this chapter is by no means a primer for those seeking a career in electric vehicle maintenance. For the mechanics who perform EV servicing, learning those skills is a specialized and multiyear discipline. The way it has been described to me is that the switch from internal combustion to electric drivetrains is more like switching from maintaining a bicycle to maintaining an airplane — they both might provide transportation but are otherwise completely different machines.

This chapter, then, provides an overview of what the average owner might expect to contribute to the EV's operation and longevity.

To summarize the topics I covered in this chapter, you'll be expected to keep tabs on these items on a somewhat regular basis:

» **Battery care:** It's a good move to become familiar with manufacturer best practices when it comes to charging and discharging the battery pack. Also, try to keep it out of overly hot or cold temperatures for long periods whenever possible.

In the world of EV's, "battery care" almost always refers to the high voltage battery pack. That said, there are two other battery-related items that may present themselves from time to time:

- *12V Battery care:* A high-voltage EV still depends on a very old-school 12V battery to power items like the headlights, turn signals, power windows, sunroof motors, and, of course, that huge touch screen display. As is the case with an ICE car, when the 12V fails, your car doesn't start. So keep an eye on said touch screen for any warning indicators, because there is no telltale slow cranking of the engine that tells you an ICE battery is about to give up the ghost. That said, the 12V batteries in EVs can be jump-started, if needed.
- *Coolant levels:* Closely related to battery care, the battery coolant helps keep your battery at its optimal temperature. Make sure you know what your model recommends in terms of a system flush — the mileage intervals can range anywhere from 150,000 miles (Chevy Bolt) to never (Model Y).

» **Tires and tire rotation:** As I mention earlier in this chapter, tires are probably the number-one routine maintenance item on an EV, if only because EVs are cars, and cars, when driven, eventually wear out the tires. Because of the weight of an EV, it's likely to happen more frequently than with a comparable ICE car.

» **Brakes and brake fluid:** Because of regenerative braking, this task might not be one that comes up during the first eight to ten years of ownership.

Because brakes stop heavy things when they're going fast, you should bring any braking issue to the attention of a certified service center at once. Hydraulic brakes will need to be flushed eventually, but this should not be a common occurrence.

» **Cabin air filters:** All cars have cabin filters and keeping yours swapped out per manufacturer schedule is another best practice.

Cabin air filters screen out air particulates like smoke, mold, pollen, and, perhaps best of all, diesel fumes from diesel-burning cars and trucks. (Those things cannot become obsolete fast enough.) So keep those items checked and changed — your lunch, and your kid's lungs, will thank you.

» **Wiper blades:** It will rain at times while you're driving, and that rain will be whooshed off the windshield so that you can see. (Even a self-driving car's cameras have to look through the windshield.) Eventually, the wiper blades will wear, crack, and generally become less effective at all that whooshing action.

Wipers usually let you know that they're approaching their golden years when they start "chattering" across the windshield. You can pop out and replace wiper blades in a matter of seconds. Read the directions on the box or pull up YouTube on your phone. I have every confidence in you.

A-a-and that's really about all there is in terms of routine electric vehicle care. Leave everything else not on that list up to the pros.

Which brings us to my favorite list in this entire chapter. Read on.

Things You'll Never Have to Add to Your To-Do List

Unless your car is a hybrid electric — either a HEV or a PHEV, which include internal combustion engines — then the following list is a helpful reminder of all you'll be leaving behind for good after you make the switch.

The items or tasks you'll never again have to factor into your schedule (and budget) include ones like these:

- Oil change
- Oil filter
- Fuel filter
- Accessory belt
- Alternator
- Spark plug
- Smog check
- Fuel injector
- Starter motor
- Muffler
- Engine air filter
- Catalytic converter

REMEMBER

Oh, and don't forget about never having to fuel the car with gas! This is an easy one to overlook, but isn't fueling up considered routine car maintenance that occurs every week or so with ICE ownership? I mean, you don't have to stop your life, drive to a specific location, and stop for 15 minutes every time your iPhone needs to be charged, do you? Of course not. If you did, that would be a huge pain in the ol' anode! My point is that going to the gas station no longer needs to be a part of your life!

I mention it at the chapter's outset, but a list of things not to do — let's call it the EV maintenance schedule *anti-pattern* — never looked so appealing.

4 Futurethink and Fun with EV Engineering

IN THIS PART . . .

Explore upcoming improvements in EV technology

Get a handle on autonomous driving

Look at other types of EVs

Take an ebike for a spin

» Figuring out where you can (legally) race your EV

» Using your EV as a backup generator for your home

» Looking at EV price trends

» Considering the role of solar and wind in the world of electric cars

» Looking at the future of fast charging

Chapter 10
Faster, Farther, and More Fun

Think of Figure 10-1 as one of those You Are Here maps:

Available data tells us that, on a global scale, humanity is still very much on the left side of the EV adoption journey, which means that the road ahead will be a steep climb up that S-curve.

And that climb will be a very exciting one, indeed.

I argue throughout much of this book that EVs are the future of personal automotive transportation, and this section does its best to look at what that future might look like.

In this chapter, I help you explore the technologies that power the electric vehicle, and what advances you might expect from those technologies in the near term,

As always, I'll be sure to point out how these advances might impact a possible buying decision today.

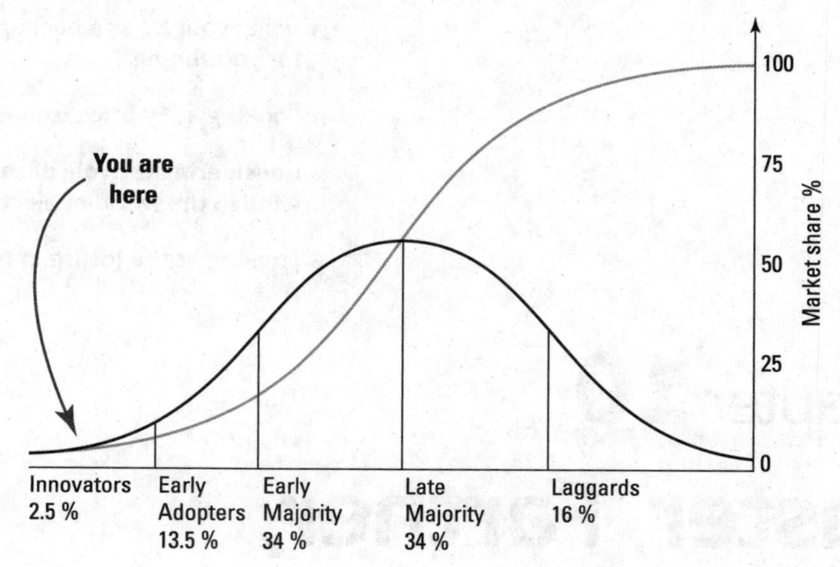

FIGURE 10-1: It's going to be a great ride.

The future is fast approaching, with all the speed of . . . well, with all the speed of an electric car. With that in mind, let's choose to ride that lightning and start our exploration.

Living Out Your Speed Racer Fantasies

I'll start things off by reiterating the fact that electric vehicles are incredibly fast modes of transportation, capable of head-snapping acceleration that no internal combustion engine can match.

But why is that? And how much faster can they go?

To understand why *electric* vehicles go so fast, it helps to first understand what's going on with the internal combustion engine.

Taking a good look at ICE power output

Imagine that you're at a stoplight in an ICE vehicle with a clear road ahead. When that ICE vehicle is at a standstill, its engine is providing very little power. Surely you've noticed that you can take your foot off the brake, and in most instances, the car moves forward. But it won't move forward *fast* because internal combustion engines have a sweet spot where they deliver maximum horsepower and torque. Usually, the sweet spot is between 4,000 and 6,000 revolutions per minute (RPMs). In fact, Figure 10-2 shows exactly what that power and torque sweet spot looks like.

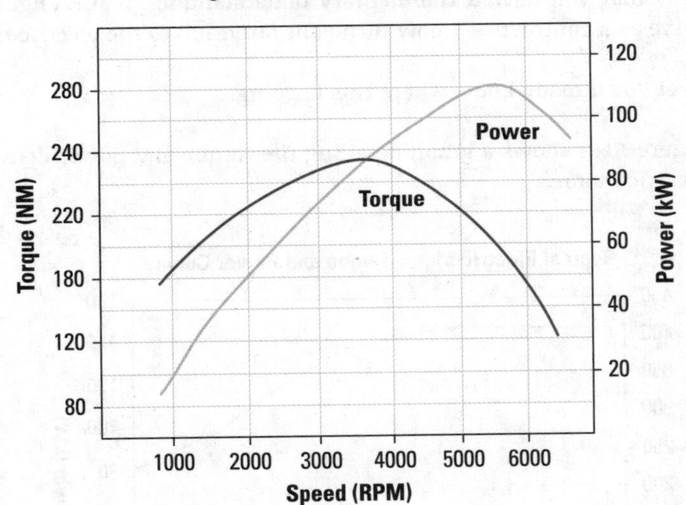

FIGURE 10-2: ICE engines are powerful, but not quick.

Fans of drag racing have observed this phenomenon dozens of times as drivers rev their engines in anticipation of the race start. At the moment the light turns green on the starter pole, there's no time to go from 1,000 to 6,000 RPMs (or whatever RPMs provide peak power for a Top Fuel dragster) — the race would be effectively over in the time it takes to reach maximum power.

Now, in theory, an ICE engine could just always operate at or about 5,000 RPMs in order to deliver maximum power every time you hit the accelerator. The problem with that, however, is that most ICE engines get extremely temperamental at 6,000 or more RPMs.

First, they get terrible gas mileage at 6,000 RPMs. Second, they tend to overheat when kept at high RPMs, which can lead to all sorts of issues, the most prominent being engine failure. To mitigate this problem, ICE engines use transmissions to match road speeds to the vehicles' most efficient RPM operating range — usually, around 2,000–3,000 RPMs.

The short version of all the info in the preceding paragraphs is that ICE vehicles possess a lag when accelerating.

Whether accelerating from a dead stop or at the speed to execute a pass, it takes time for that engine to rev to its peak power output RPMs.

Comparing EV power output

Now that you have a rudimentary understanding of ICE engines — enough to serve as a contrast — I now turn your attention to the electric motor.

I bet you already know where this is going.

Figure 10-3 shows a graph detailing the torque and power delivered by a typical electric motor.

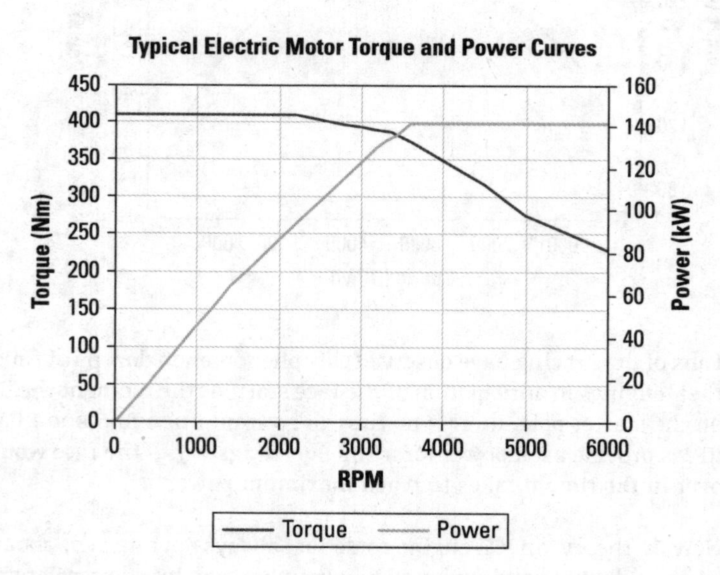

FIGURE 10-3: EV motors are powerful *and* quick.

Ok. So I've just presented two charts with some squiggly lines. So what?

What's most significant about the second set of squiggles is the torque line, which starts off at peak and doesn't begin to drop until the motor is cycling at 3,000 RPMs.

What that line means is that the power plant in an electric vehicle delivers almost maximal output at zero RPM. In addition, EVs typically don't require a transmission in order to optimize acceleration, as is the case with an ICE car.

What's more, because EV motors have such a small footprint when compared to their ICE counterparts, manufacturers of high-performance EVs — think Tesla Performance models, Lucid Airs, Porsche Taycans, and the like — typically include one motor at each axle. One motor per axle means more traction for the tires, which in turn can channel all that torque and power into the pavement rather than simply spin the wheels.

The weight of the battery helps keep the wheels glued to the ground as well.

How this is translated to the real-world driving experience is that even the slowest electric vehicles are incredibly quick from a dead standstill.

Almost *any* electric vehicle — from a $40,000 Hyundai Ioniq5 or Chevy Bolt hatchback to a $50,000 BMW i3 to a $165,000 Lucid Air Dream Edition — is quicker off the line than an ICE counterpart and then some, whether we're talking about a $50,000 Chevy Camaro, a $100,000 Chevy Corvette, or even a $300,000 Ferrari F8 Spider supercar.

REMEMBER

I've driven several electric vehicles in full "launch mode," which creates the conditions necessary for the very fastest 0-60 times possible by cars like the Model S, Taycan, and the Air. A sub 3-seocnd 0-60 time is so fast that it feels like teleportation.

It's like I half expect to step out the car only to discover that my kids are now older than me.

You can thank the torque squiggle in Figure 10-3 for that.

EVs are also head-snappingly quick when punching the gas during normal traveling speeds, even on a highway. Why? Because EV motors don't have to wait for the transmission to downshift one or more gears in order to deliver that peak power.

Yes, I know: In a full quarter-mile drag race, the Ferrari would handily beat the Hyundai or the BMW — here's where that whole gearing thing comes in handy — but these just aren't the things that 99.9 percent of the driving public cares about.

Then again, you might be the owner of the Lucid Air. (And by you, I mean me.)

Or, you might own a Model S Plaid or a Porsche Taycan Turbo. Or, heck, you might even own the electric Lamborghini Terzo Millennio. (See Figure 10-4; and by *you*, I most definitely *don't* mean me.) Quarter-mile track times might not be top of mind when driving to your kid's school play or swim meet, but then again, car enthusiasts buy powerful cars for a reason, one that has everything to do with car enthusiasm, and absolutely zero to do with a midlife crisis, or at least this is what I'm told by a friend.

FIGURE 10-4: It's a concept car, so no one may even own one. But still.

Mike Mareen / Adobe Stock

In any event, where exactly does one take an incredibly powerful car, electric or otherwise, to truly experience the limits of the powerful motors within? Asking for a friend.

TIP

It's like the advice you may have gotten before going for a swim: Be careful about eating right before. I've known people who have gotten out of an EV after several test drive launches and promptly launched their last meal onto the pavement.

I know. Awesome, right!?

Staying Out of Jail

Let's say that you own, or otherwise have managed to obtain for a day, a conspicuously fast electric vehicle, and you're also the type who's thirsting for, or thirst-Tweeting about, an experience that answers any or all of these questions:

- Just how fast can this thing go?
- What would win in a ¼-mile drag race — a silver Aston Martin DB9 or a stainless steel Cybertruck?

» And by *fast*, are we talking about the first drop of a roller-coaster fast? Or are we talking about CERN particle-collider fast?

» What is all this Nürburgring stuff about, anyway? Should that be a thing I care about?

» Did I miss my life's true calling as a race car driver? Am I going to deeply regret my golden years, knowing that I was *this* close to a life measured against the likes of racing greats like Mario Andretti, Michael Shumacher, or Ricky Bobby?

It may not surprise you to learn that the answer to these questions, along with many others, can be found only on a race car track.

Fortunately, it's *relatively* easy to arrange to drive your car — EV or otherwise — around a track. All it really takes are 1) a track and 2) money. Let's look at four places around the continental US — and one that's not — that are open for business to weekend racing enthusiasts:

» **Sonoma Raceway, Sonoma, CA:** I'll start with the one that's right in my back yard. The Sonoma Raceway is a 12-turn, 2.53-mile course that's nestled into a hillside along the southern edge of California's Sonoma County.

  The drag strip is open to all on Wednesday nights from March through November. (See Figure 10-5.) And — hey! — if you're visiting with your Tesla S or Porsche Taycan in tow, drop me a line and I'll recommend a good winery in the area (there's only about 3 gabillion to choose from).

  Oh, and I'm not saying that, were you so inclined, you could use a trip to California wine country as an excuse to plan a trip out here and then just *happen* to notice a racetrack nearby so you might as well check it out, hon, but I'm not *not* saying that, either.

FIGURE 10-5: I'm not saying that's me drifting a Lucid Air. I'm not *not* saying that, either. (Yes I am. I'm saying that wasn't me in that car.)

Lucid Motors

- **Lime Rock Park, Lakeville, CT:** Those in the northeast US might consider the 1.5-mile, 7-turn Lime Rock racetrack that's tucked into the woods of northwest Connecticut. No straight-line launches allowed, though. Racing here means first signing up to become a Lime Rock Drivers Club member, which goes for about $1,500 at the time of this writing. Kind of a steep price tag, but then again, performance EVs aren't exactly free, either.
- **Circuit of the Americas, Austin TX:** Why do I have a feeling that the Circuit of the Americas will get a lot more popular with weekend EV racers sometime around March of 2022? Oh yeah, it's because Circuit of the Americas is in Austin, and so is Tesla's Giga Austin. The track consists of a 3.4-mile circuit containing no fewer than 20 turns — there's a reason that this track is the only one in the US that hosts a Formula 1 race.
- **Road Atlanta, Braselton, Georgia:** In recent years, there's been a lot of tech jobs and real estate development and just general cash sloshing around Hotlanta like so much sweet iced tea, and about an hour from downtown you can find a 2.5-mile, 12-turn course where you can book some time to go out and flex your car's electrons.

  There's an event page for Riad Atlanta at `www.roadatlanta.com/events/track-days-car-clubs/view-track-days-schedule`, and the advice is to hook up with one of the car clubs that regularly books time at the track.
- **Nürburgring, Nurburg, Germany:** The Nürburgring (see Figure 10-6) is perhaps the most famous racetrack in the world, although technically it's two tracks — the 5-kilometer Grand Prix (where they hold Formula 1 races) and its older, better-looking, and much more famous brother, the Nordschleife.

  The Nordschleife, also known as the Green Hell, is the track that so many EV automakers make so much fuss about. The general consensus is that it is, and has always been, one of the most beautiful and challenging race tracks on Earth.

TIP

You can get married at the Nürburgring, if that's your thing. Personally, I'd prefer a trip to California wine country and then dinner at The Girl and the Fig in downtown Sonoma, but to each their own.

In any event, the point is that you, too, can race the Nürburgring using your own car. As you've seen, you can pull off this feat in California or in Texas or just about anywhere — or you can race at the world's most revered raceway. Right there on the website, under the Driving tab, is this heading:

> Experience the Nürburgring In Your Own Vehicle

If you're in the area and you have a car handy, you can do worse than browse to `www.nuerburgring.de/driving/touristdrives`.

FIGURE 10-6: Hell isn't so bad, after all.

Markus Volk / Adobe Stock

TECHNICAL STUFF

The EV record for the Nürburgring Nordschleife is held by the Tesla Model S Plaid, which beat the Porsche Taycan's previous record by 11 seconds. Rest easy, Porsche fans: The overall lap record is held by a 911 GT2.

Can an EV Power Your House?

Yes, it can.

However, *can* is doing a lot of work in that previous sentence (and heading!), so let's talk about this further.

On the surface, powering your house with your electric vehicle is a simple concept. As you (now) know, a big battery like the one in your EV can store enough energy to power a 3,500 lb. vehicle for 300 miles, or sometimes more. Surely that same battery could also power a refrigerator for a while during an outage. Days, even.

Indeed, enough juice is stored in the typical EV battery pack to power an entire *house* for several days if that house loses grid power.

Or, as an alternative to this power-in-case-of-emergency scenario, your EV could charge to (almost) full during off-peak times and then supply that stored power back to your house during peak evening hours, saving you hundreds every year on your utility bill. (If you don't have a time-of-use power plan, you can ignore this paragraph. Those who do have such a plan know that they pay much less for electricity during the night than they do from 4-9PM.)

As a bonus, if you and several of your neighbors adopted this habit, it would reduce the strain on the overall power grid.

Vehicle-to-grid usage also has the potential to not only leverage cheaper power generation but *cleaner* generation as well.

Generally, daytime electricity generation can leverage sources like solar and wind more readily than nighttime generation, which typically rely on coal or gas-fired power plants.

What I've just described is something called vehicle-to-home (V2H), as in the vehicle supplies power to the electrical grid that normally powers your home.

And, if your vehicle could store power — maybe purchased at a discount during off-peak times, or even more ideally from a solar array on a homeowner's roof — and then supply that power back to the broader electric grid, then you have a solution called vehicle to grid (V2G). V2G would let an EV owner turn a solar array/EV combination into a mini power plant that generates both electricity and profits for the owner.

But — does it actually work?

No. It doesn't.

Or at least not yet. Let's look at some of the reasons why.

Recognizing the challenges of V2H

For starters, a nonzero number of EV manufacturers have decided that this V2H business just isn't worth devoting time and resources to for now. Tesla and Chevy, for example, aren't making any vehicles with V2H or V2G capabilities.

Not only are there significant hardware and software implications just for the car, also keep in mind that Tesla envisions an autonomous future, which I explore further in Chapter 11.

If self-driving cars were constantly on the road, they wouldn't be available very often for grid use. In theory, every time an autonomous car plugs in, it would be to draw power from the grid, not the other way around.

For V2G or V2H to work, the functionality has to be built into the vehicle. (Tesla does make the Powerwall, which does indeed power your house. It's effectively an EV battery, just one bolted to the wall rather than to a car chassis.) And because we're talking about the makers of the Bolt and the Models S3X and Y, we're talking about millions of EVs without V2H capabilities.

So, for now, it's up to other manufacturers — like Lucid with its Air sedan, Ford with its Mach E and Lightning F150, and even Nissan with its newer Leafs — to get the ball rolling.

And yet — V2G and V2H capabilities are still not available on many of those cars/trucks, either (except in very specific circumstances). Right now, even for those pursuing this goal, they exist mainly in product roadmaps and marketing campaigns.

Why not? Or why not now? In a sentence, it's because a lot of houses aren't wired for it.

To understand why, consider that the electric grid is just a vast network of interconnected wires. If you live on a typical suburban US street, the houses on that block are all essentially connected to a length of copper wire. (It's why a power outage usually affects many houses at a time.)

For a V2H system to work, then, there has to be a means of disconnecting the breaker panel — the house — from that common copper wire. Why? Lots of reasons, and one of the foremost is that doing so protects workers trying to get the grid functioning again from being shocked by a big EV battery pumping electricity onto that wire.

In other words, you need a way to make sure the juice from the car battery is isolated to just that house, and the vast majority of breaker boxes just aren't wired this way.

REMEMBER

If you have rooftop solar, it's likely that your breaker box *is* wired exactly this way — the way that allows a two-way flow of electricity. Check with your favorite electrician of course, but a V2H setup should prove much easier to implement on a home that is using rooftop solar.

Adding the challenges of V2G to the mix

The main challenge to the advent of ubiquitous V2G can be found in the very description of the technology. Again, this is something that potentially would allow someone with a solar array and an EV to sell electricity to the power company — at a profit.

See the challenge here?

Over 40 states allow *net metering*, which is a system that connects power generators (like a rooftop solar array or a battery) to a public utility grid in order to transfer excess electricity onto the grid. Utility companies, however, are not states and so just because the state allows it doesn't mean it will be feasible with the electric company you use.

Utility companies are in the business of generating and selling electricity, after all. This is a very heavily regulated industry anyway (as it should be), so it's not like V2G is something that would be implemented overnight, even if lots of utility executives were enthusiastically behind the idea. There would be many layers of state and local debate, approval, and implementation oversight. That may read as "red tape," but for the most part, red tape provides a tremendous societal good — for the most part, red tape keeps the public safe rather than sorry.

REMEMBER

The current coming out of a car battery would be direct current (DC) output. It must be converted to alternating current (AC) for use by either a home or the grid.

For vehicles that have this capability, there's an inverter that handles this task. Another option would be to build the inverter into the wall charger, also referred to as electric vehicle supply equipment (EVSE). In short, using V2G would certainly require car hardware and software that would allow two-way flows of electricity, and also *may* require the kinds of dedicated wall chargers mentioned in Chapter 7.

Besides, utility executives are *not* enthusiastically behind the idea. They're enthusiastically behind the idea of selling you electricity. They're not enthusiastically behind the idea of ensuring that infrastructure is in place so that you can sell it to them. In fact, my educated guess is that the only way there will be widespread adoption of V2G (or V2H) is if those same state and local governments compel utility companies to do so.

In short, my opinion is that if you want V2G, buy a Ford or Lucid or Nissan, yes, but also run for city council.

Addressing challenges with a Smart Grid

The innovation that is perhaps most needed to make V2H and V2G possible doesn't rest with the vehicles. It's the grid.

Imagine you're a utility executive in Portugal or Chile or South Africa and you've just made a large investment in a solar array. Your utility company will generate thousands of kilowatt-hours' worth of electricity, and now you'll need something on the order of 50 gigawatt-hours of battery storage so that all the solar energy generated during the day doesn't go to waste once the sun goes down.

Fifty gigawatt-hours will be a very expensive battery.

But now imagine that your country has over a million electric vehicles on the road, each with at least a 50-kilowatt hour battery pack. More batteries, sure, but the same amount of potential energy storage — 50 gigawatt-hours.

If only there were a way to utilize all those individual batteries in a way that was good for the grid, and good for the individual EV owner.

The answer, my friend, is software. The answer is the smart grid.

In fact, my home state of California *already* has more than 1 million electric cars. One million EVs on the roads and, more significantly, parked in offices and homes represents a potential virtual power plant that could be used to all but eliminate California's need for fossil fuel electric generation.

The missing link for the VPP is what's called a smart grid — software, in other words.

With a smart grid, utility companies can use the Internet to first access batteries and then manage the flow of electrons in real time, balancing the needs of the grid and charging demand. Here's how this would work:

1. You're at the office, connected to a charger all day, and for most of the day, your car is charging up from 50 percent to 85 percent. Thanks, California photons!
2. As the sun begins to set and you're still at the office, the smart grid detects higher demand, so your car battery sends power back to the grid during this demand spike. (Don't worry: The software makes sure you're compensated for that power.)
3. You leave the office with your car's state of charge back to 50 percent, but that's more than plenty to make it the ten miles back to your house and even back again the next day. And of course, the smart grid software would include the ability to set exceptions.

   Say it's a Friday and you need the car fully charged to start a weekend road trip to wine country. You can use an app to tell the grid not to draw from your battery that afternoon.

The net gain of all these smart grid machinations is the creation of one big distributed battery system rather than one utility-scale system only for utility companies, and one distributed personal-scale system only for EV use.

If we can get the smart grid implemented, it would allow each battery in circulation to pull doble duty, thus dramatically reducing the strain being put on the battery supply chain. As it is, the supply chain is struggling to supply the planet with enough batteries to meet demand.

Alas, no such smart grid exists today, or at least not on the scale of something like the state of California. (PG&E claims that it will begin some trials in 2022, but PG&E has made this claim previously.)

But, as the saying goes, software is eating the world, and here's to the day when it eats significantly into humanity's need for batteries. It'll be incredibly difficult to implement at scale, but as I pointed out in this book's Intro, humans are capable of amazing ingenuity.

We put men on the moon before we invented the cell phone. We can figure this out, too.

And, while we're waiting for that day to arrive, we can use that time to ponder the question in the following heading.

Will EV Prices Ever Come Down? When?

Um . . . not really.

If you're waiting for the price of an electric car model you've got our eye on to come down in price, I'd recommend either casting your eye in another direction, or resigning yourself to stretching the budget just a bit when getting your first EV.

That's just the way things have been for the past four years, and I don't think conditions will change much in the next four years. Demand should continue to rise faster than cars can be made, which in turn will put inflationary pressure on the raw materials necessary for EV production.

Yes, we might continue seeing declines in the battery cost curve, eventually reaching that magical $100/kWh inflection point, but by that time, the price of everything else may well have headed in the other direction — tires, glass, seating, steel, silicon chips, and the salaries of software engineers may have all significantly risen by the time batteries hit $100/kWh. (Believe it or not, I wrote that paragraph long before the 8% inflation rates of mid-2022.)

What I'm saying is that the $35,000 Tesla Model 3 was available for about 18 months in 2019–2020 (and even then, it was an off-menu selection); a new

Model 3 will *never* be available at that price again. (Current starting price is over $47k.)

What I'm really saying is that the best time to buy an electric, including the Model 3, was a year ago. The second best time is today.

Speaking of: I'm here to take all bets that the $39,000 Cybertrucks or F-150 Lightnings will *never* functionally see the light of day. (I'll allow for a few rolling around at a steep loss to the manufacturer just for the bragging rights). In the parlance of sports gambling, I'll take the over, as in those things will be a lot more than $39,000.

According to the folks who engineer these things — people I deal with every day at work, as it so happens — a truck at that price is simply not possible. Unless there's some radical shift of raw material cost, the $39,000 truck is either a pipe dream or a marketing stunt — take your pick. So be mentally ready to ante up more like $80,000 for that electric pickup.

No one said saving the planet would be free.

REMEMBER

EV truck owners stand to gain even more than their sedan- and SUV-driving brethren when it comes to monthly fuel savings. A typical pickup gets about 15 MPG. The Rivian R1T's MPGe is 70, which means hundreds in fuel savings every month.

If you're looking for less expensive electric transportation in the short term while still buying new, here are some options/models worth investigating:

- Chevrolet Bolt (from $25,600)
- Nissan Leaf (from $27,400)
- Hyundai Kona (from $34,000)
- Kandi K23 (from $20,000)
- Mini Cooper SE (from $30,750)
- Kandi K27 (from $11,999)

REMEMBER

In the US, the federal plug-in tax credit still applies to most of these vehicles, reducing the price by a further $7,500. And state and local incentives could further reduce the costs of *any* of these models.

The point is that except for the two Kandi models, not a single one of these new EVs is what I'd consider an inexpensive car. But neither are cars — period. Heck, even the Kandi is over five times more expensive than what I paid for my first car.

Then again, I purchased my first car in 1987, and even an equivalent of that car today would run about five times more than what I paid back then.

I mean, pick a consumer product, and then think about whether it has dropped in price over the past five years or so. Bacon and other staples? Nope. Movie tickets? Nope. Phones? Hardly. Running shoes? I'm detecting a pattern here. Houses? Ha! Ha-ha!! This is me, using two exclamation marks to indicate that I'm laughing manically at that last one.

Are there exceptions? Of course. But as a rule, I wouldn't count on big price declines for EVs anytime soon. (And it bears repeating that the average new car in the US now goes for over $45k.)

Renewable Energy and the Electric Car World

Now then: Consider Texas.

Consider a state whose residents elect politicians who crusade against gub'mint regulation, get elected to office to legislate in harmony with that crusade, and, once there, deregulate the electricity industry to such an extent that there are now 150 electric providers operating in Texas. One hundred. And fifty.

Apparently, anyone who can slap together a WordPress website is qualified to deliver power to banks, hospitals, schools, factories, and other buildings. But hey, it's not like electricity plays any essential role in a functioning society.

And yes, the preceding paragraph may sound ridiculous to every sentient being who's *not* serving a term in the Texas legislature, but remember: Average Texans have been rewarded for their deregulation proclivity by . . . wait for it . . . *paying almost $30 billion more* for power than people in other states over the past 15 years! Three cheers for deregulation! (I invite you to check out the supporting article at this *Wall Street Journal* page: `www.wsj.com/articles/texas-electric-bills-were-28-billion-higher-under-deregulation-11614162780`.)

But really, I say consider Texas because Texas has three things in abundance:

- Sunlight
- Wind

» Politicians who fail their constituents through deregulation, only to keep getting reelected so they can repeat the cycle

Yes, consider Texas, indeed.

The good news, though, is that only two of those three things are needed to power your electric vehicle with 100 percent clean, renewable electricity.

The even better news is that from almost anywhere in the United States and in many other countries, you can opt for this clean, 100 percent renewable energy even if you're hundreds of miles away from the nearest wind or solar installation. (You can even do this if you live in Texas, although I feel a little like this is rewarding bad behavior.)

How?

As it happens, it's all about *who* collects your utility check.

Using a clean energy broker

Lots of people can procure renewable energy for their home even if their local utility doesn't generate electricity that way. How? By using a broker.

Energy brokers act as a mediator between energy suppliers and their clients. These brokers don't distribute energy, they don't own energy, and they can't sell energy directly to you. They simply present the rates of a wholesaler or supplier. In my case, the broker ensures that my utility bill is used to pay for renewable energy rather than energy generated by fossil fuels.

In other words, my residential energy broker (Arcadia Power) takes the dollars I pay to Pacific Gas and Electric, pays them for my energy use, then uses my dollars to buy wind energy generated by wind turbines on the wide-open plains of Texas. (I further assume, but don't care enough to investigate it, that the broker pays PG&E wholesale rates, and then likewise pays wholesale to Texas. There's got to be some way Arcadia makes a profit.)

Several utility brokers are now operating across the United States, Canada, and Europe. Here are a few that are worth a look.

» **Arcadia Power:** Signing up with Arcadia is quick and easy. The way it works is that it bills you what your electric utility bills you anyway, and they pay the utility on your behalf. In return, Arcadia takes your money and buys power from renewable sources. As it happens, all the electricity in my home and in my cars comes from a wind farm in Texas, so maybe I should give poor Texas a break. (See Figure 10-7.)

I've been an Arcadia customer since 2018, before my family got our first EV, and so from the very first day of ownership, we've been powering our car with 100 percent renewably sourced electricity. And that's pretty cool.

Find out more about Arcadia at `www.arcadia.com/about-us`.

And you certainly don't have to, but you can tell them a friend sent you: `https://arcadia.com/referral?promo=brian29929`. In fact, if you email me confirmation that you've signed up, I'll send you an autographed copy of the book. (My contact info is in the front of this book.)

- **Rhythm Energy:** Another company that offers consumers the option to purchase 100 percent renewable energy is Rhythm Energy, an outfit based in . . . Houston, Texas. Wait — what? Why does this chapter keep windmilling its way back to Texas?

  Yet it's true: Rhythm Energy lets Texans choose from plans like the Simply Green plan, which delivers electricity from 100 percent clean sources like these:

  - Solar farms in Texas
  - Wind from the Great Plains (which also includes Texas — I mean, have you been to Amarillo?)
  - Hydro on the west coast

FIGURE 10-7: Who doesn't love Texas?!

The biggest difference between a company like Arcadia and one like Rhythm is that Rhythm is the actual utility of record. But functionally, it's all the same to the end user. You pay someone and you get electricity. The upside of using one of these providers is that it's a way for you to have some control over how that electricity was generated.

TIP

Rhythm also has a very cool EV ownership calculator, found here: `www.gotrhythm.com/ev`. Just punch in a gasoline vehicle and a comparable EV and see where things land. For example, when comparing the total cost of a BMW 330i with a Tesla Model 3, the 3 costs about $4,500 less over five years than the beamer.

» **Google:** Yes, even Google can help you power your EV with cleaner grid power, through its Nest Renew program. This is a brand-new program at the time of this writing — and a popular one because there's a waitlist to join — so I'll just mention it and leave a link where you can find out more or get in line: `https://nestrenew.google.com/welcome`.

It's free to join for Nest thermostat users, but for me, the most significant aspect worth mentioning is that, like Arcadia, Google's Nest Renew offers a $10/month add-on that supports wind and solar projects.

Using your own solar power

Having an electric car and then developing a keen awareness of where that electricity comes from almost go hand in hand.

And what better electricity is there for your car than the kind you can harvest from your own rooftop?

Now, I won't go into any detail about solar installation, as the subject is one for another book, but what I will say is that the cheapest way to refuel your car's battery is to generate that power yourself using a solar energy system. The cost for each kilowatt hour?

Free.

Yes, there exist significant up-front costs when installing home solar, so really it's not free at all; it's just that there's no marginal cost for generating electric once the system is up and running.

REMEMBER

If you drive 12,000 miles a year, you'll need about 3500 kWh to power a single EV. You'll want a 2–5 kWh solar panel system to cover that amount.

Yes, there are tax credits and other incentives available from state governments, and even sometime utilities. Yes, there are many companies that will help you get set up, although waiting times can be long indeed. Yes, you may even be able to sell excess electricity back to the grid, helping offset your initial capital expenditure.

Starting in 2020, all new single-family homes, multi-family homes up to three stories high, and commercial buildings *must* include solar panels. So says the law in California. Way to go, California legislature!

Now if they could just abolish prop 13 . . . (It's a property tax thing that keeps housing prices artificially high.)

When all costs are factored in, you can estimate that solar charging of your EV will save you about 50 percent versus getting that power from your local utility. Of course, your mileage will vary depending on the size of the system and how long you stay in your home.

Some energy brokers like Arcadia Power let you invest in community solar, where you buy solar panels that aren't installed on your roof, but rather in solar farms located somewhere else. This can be an ideal solution for renters or those who otherwise can't control what goes on the roof where they live. The investment in community solar pays you back with a reduction in your monthly electric bill. Unfortunately, not all states allow community solar. Silly, I know.

Disclaimer: I do not own solar. There are two YouTubers worth a follow who do, however, and have made videos about their experiences at the 4-year and 10-year mark of solar panel ownership. If you're considering solar, I highly recommend giving these a few minutes of your time:

- **Two Bit DaVinci's report after 10 years of solar:** `https://youtu.be/PXk5AFH2LaY`
- **Undecided with Matt Farrell after 4 years:** `https://youtu.be/jxt3v7kTlvw`

Next-Gen Batteries

The topic of next-gen batteries is very broad — something that could easily fill its own book and something that certainly *does* fill several semesters of engineering study.

First, know that the term *next-gen* means whatever technology *seems* like it's just a few years away from utterly revolutionizing the market for energy storage. There have been next-gen batteries for as long as Alessandro Volta invented the voltaic pile back in 1800. (See Figure 10-8 for a look at Volta's innovation.)

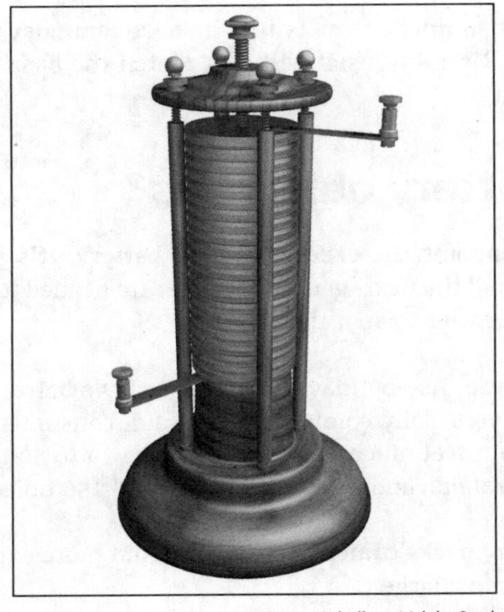

FIGURE 10-8: Believe it or not, this thing stores electricity. Believe it or not, it will not power an electric vehicle.

J J Osuna Caballero / Adobe Stock

Second, there are people, like University of Texas professor and Nobel laureate John Bannister Goodenough (yes, his name is Johnny B. Good . . . enough), whose entire professional existence is dedicated to next-gen battery research.

And he's one of the best who's ever lived. Among other career accomplishments, Goodenough is widely recognized as the driving force behind the lithium-ion battery, the very thing that, along with Nikola Tesla's induction motor, has enabled the electric-car revolution.

So, yeah. People like him have been defining next-generation batteries for several decades now. My task is to present, in the space of a few paragraphs, an overview about what the likely near-term future might look like. I'll do my best, but please remember that I don't have a crystal ball, nor do I have access to the research laboratories of battery chemists. I only know what others report about what Goodenough and his ilk are cooking up in his UT digs.

TIP

Oh, and get this: John Goodenough won his Nobel prize at age 97. No, that is not a typo, and yes, that makes Goodenough perhaps the most badass human walking the planet. Ninety-seven.

Sometimes at dinner parties, I humble-brag that I've written 20 or so technical books in my life, but, hey ladies, who's counting? Meanwhile, Goodenough eats 20 books for breakfast and then probably goes outside to, I don't know, perform feats of strength like bench pressing Bevo, Texas University's 2,000-pound. longhorn-steer mascot.

That paragraph is not so much a tip as it is a sober reminder that neither you, nor I, nor even Chuck Norris himself will ever match the badassery of one John Bannister Goodenough.

Next-gen battery objectives

Rather than try to pinpoint the exact next-gen battery, it's perhaps wiser to describe the goals that all the next-gen EV batteries are headed toward. The *what's* of next-gen are much easier to spot than the *how's*.

Let me start with a recap: As you may have read, today's battery packs are more than enough to cover your daily commute. That said, consumers generally want their EVs to be able to travel much more than a mere 30 to 50 miles; consumers are voting with their wallets, and their wallets demand 250 miles, at a minimum.

But, with bigger battery packs comes heavier cars. And more expensive cars. And packs that take longer to charge.

And thus we arrive at the three main pain points of current battery technology, and thus we arrive at the three predominant objectives of any next-gen battery: Scientists and companies across the globe are engaged in an ongoing quest to make batteries

- Lighter
- Less expensive
- Faster charging

Let's look at where we're heading on each of these three fronts.

Less expensive (and faster charging)

This is not news to most of the planet, but a Chinese manufacturer by the name of CATL is the world's largest supplier of EV (read: lithium-ion) batteries.

And, in late 2021, CATL — an abbreviation of Contemporary Amperex Technology, Ltd, by the way — announced a bold plan to eventually replace lithium-ion business in favor of next-gen sodium-ion (Na-ion) batteries instead.

Why would they do that?

The primary advantage of sodium-ion chemistry, as of this writing, is one of cost. Sodium-ion batteries don't use rare metals (not to be confused with rare-earth metals), so obtaining the raw materials needed for battery production is more cost-effective and more cost-predictable.

For example, even though China is one of the world's largest producers of lithium carbonate (the stuff one needs to make lithium-ion batteries), the nation still must import from other top producers like South America and Australia to meet growing demand.

Unfortunately for those in the battery cell business, lithium prices are subject to wild price swings. For example, in 2021 the price of lithium carbonate more than doubled when compared to 2020 prices.

On the other hand, to mine the sodium needed for sodium-ion batteries, all one needs is seawater. (Technically, you need to evaporate the seawater and then scoop up the salt that remains.)

Another advantage of the sodium-ion technology is that these Na-ion cells recharge faster than their Li-ion counterparts. And, according to CATL, they perform better at low temperatures. (Read all about it at `www.catl.com/en/news/665.html`.)

It's not all wine and roses, though. One drawback, for now, is that the energy density of sodium-ion batteries is not yet up to par with lithium-ion cells. Or, in other words, if there are two 75 kWh battery packs using these competing chemistries, the Li-ion pack will be smaller than the Na-ion. (As you might have guessed, CATL researchers are hard at work trying to make Na-ion densities match that of lithium.)

One dummy's verdict:

I give Na-ion a 9-in-10 chance of coming to market. No, that verdict isn't worth the paper it's printed on, but CATL is the biggest supplier, after all, and if the biggest supplier decides to change direction, auto manufacturers that rely on CATL batteries really won't have much choice.

REMEMBER

It's not just CATL. About 30 other companies around the world are pursuing sodium-ion battery solutions, and 30 companies wouldn't take such a shared risk if there wasn't some *there* there.

Solid-state batteries: Lighter weight (and faster charging)

Solid-state batteries have been a darling of next-gen battery discussions for as long as I've been driving and learning about electric cars — so, about five years.

Solid-state batteries get their name because they don't use a liquid electrolyte; instead, they use a solid electrolyte to separate anode from cathode. But the reason they're called solid-state misses the larger point, which is about the promise they hold for EVs.

Once the topic is introduced, it's easy to see why folks get so excited: Solid-state batteries exist today, and they have an energy density that's at least *two times* greater than lithium-ion batteries. In addition, solid-state batteries are safer because they don't use toxic, flammable liquid electrolytes. In turn, this makes the solid-state batter more efficient because you can devote more of your battery real estate to anodes and cathodes rather than to components or structures that keep the battery safe.

In short, solid-state batteries mean smaller battery packs, which mean lighter cars, which mean more range, all of which could theoretically lower the overall price. (More about that in just a moment.)

Plus, because you don't have to worry about heat as much as you do with liquid electrolyte batteries, you can charge them faster. (In lab tests, solid-state batteries charge in about $\frac{1}{10}$ the time it takes for conventional Li-ion batteries.) Oh, and did I mention no more battery fires?

I mean, what's not to love? Solid-state and EVs are a match not seen since chocolate and peanut butter first hooked up all those years ago.

What's the catch? They're great in every way, and we know how to make them, so why aren't solid-state batteries already the standard in today's electric vehicles?

Alas, two major factors are stalling the revolution. (Note to self: Use a different gerund. "Stalling" is a terrible analogy in a book about EVs. Unlike ICE engines, electric motors don't stall.)

The first factor is price: The economies of scale aren't yet in place, making the cost of an EV-size solid-state battery pack cost prohibitive for now.

The reason solid-state batteries are expensive brings us to the second drawback: They're difficult to manufacture. Oftentimes, mass production results in tiny amounts of gas being trapped between the solid electrolyte and the cathodes/anodes they're separating. This can cause major problems with battery performance — without making this a science lesson, bad things happen when trapped gases touch other things inside a battery housing.

One dummy's verdict:

I give solid-state a 4-in-10 chance of coming to market in the next three years.

Big battery manufacturers, like Samsung, through its Advanced Institute of Technology, continue making progress on solid-state battery technology with each passing day.

And Fisker announced in 2017 that its first car — to be released in 2023 — would use solid-state batteries.

The problem is that Samsung's solid-state batteries remain in the lab and not rolling off the factory floor. And you did catch the date of Fisker's announcement, yes? Yes.

The industry has been touting solid-state for a long time now with little to show for all the touting. What's more, Fisker *shelved* the solid-state battery plan soon after the announcement, opting instead for a traditional Li-ion battery pack. (And this is if Fisker ever makes a car, which I also see as a remote possibility. For the backstory on this one, check out `https://electrek.co/2017/11/14/fisker-solid-state-battery-breakthrough-electric-cars`.)

Lithium-sulphur: Lighter, denser, cheaper

One last item to keep an eye out for is another chemistry change: lithium-sulfur.

There's no question about the advantages offered by lithium-sulfur (Li-S) batteries. In a word, batteries using this chemistry are lighter and cheaper to produce. As a feather in the cap, these batteries can store more energy in an equal-volume battery than can lithium-ion.

The higher energy density means that the Li-S battery can last longer on a single charge. Additionally, lithium-ion batteries use cobalt — a metal that, like nickel, is subject to wide swings in price, depending on global supply chains — whereas Li-S batteries use sulfur, which is a plentiful raw material, albeit in part because it's a common byproduct of oil production. The fact that sulfur is abundant and cheap is the main reason that Li-S batteries offer a cost advantage. And the news gets even better: Lithium-sulfur batteries can be mass produced on the same equipment used for Li-ion manufacturing, so a switch from one battery chemistry to another should be relatively painless.

If you find yourself squinting right now because you're thinking, "Hey, Brian, this is all starting to sound a little too much like the solid-state story," then imagine me offering you a virtual Nancy Drew decoder ring as reward for your ability as gumshoe.

Because the other shoe, as it were, is that lithium-sulfur batteries don't last very long. Certainly not long enough for cars.

For the moment, the reason this chemistry isn't already playing a starring role in your phones and EVs is that they can't be recharged enough times before they fail. Current Li-S batteries only make it to about 50 recharging cycles before severe degradation or even shorting. This is obviously nowhere close to the number of recharge cycles needed for a commercially viable laptop battery, much less an electric vehicle pack that's expected to keep a majority of its capacity for many hundreds, or even thousands of cycles.

Now, why this happens is fairly well-known, and scientists are hard at work addressing the cause, and blah, blah, blah.

One dummy's verdict:

I give lithium-sulfur batteries a 2-in-10 chance of coming to the EV market in the next five years. I do believe, however, that some of the work being done to improve Li-S batteries will be used to improve lithium-ion technology. I also believe that these batteries will continue to see use where they are already deployed: in devices that need lightweight, long-lasting batteries, like satellites and drones.

Faster charging (that is all, but that is also enough)

Another interesting next-gen battery takes quite a different approach. Rather than improve the battery chemistry (as with Na-ion), or improve the form factor (solid-state), some folks are trying to improve the existing lithium-ion batteries with materials that allow them to charge much, much faster than what's possible today.

Exhibit A in this approach is a company called Battery Streak. Battery Streak uses "nanostructured materials to dramatically decrease lithium-ion intercalation distance in the charge process." (At least that's what they say at `https://batterystreak.com/tech`.)

Oh, sure. That old intercalation thing. Who *hasn't* dealt with that?

In more everyday English, it means that whenever a traditional Li-ion battery recharges, lithium ions are moved from anode to cathode, although on a microscopic level, the process isn't as efficient as it could be. Essentially, a lot of lithium ions get "stuck" at the front entrance of the cathode. Imagine a crowded theater, perhaps, where everyone heads for the exits at one time and there's a pile-up of humans where they have a chance to leave one room and disperse into another.

Battery Streak's use of nanostructured materials means that the cathode presents a more porous surface for the ions to pass through, resulting in a more efficient — and much faster — charge time that also produces much less heat as a side benefit. Imagine the same theater, only with dozens more exits.

Another entry in its assets column is that the company "has been awarded a National Science Foundation (NSF) Small Business Innovation Research (SBIR) Small Business Technology Transfer (STTR) grant to conduct research and development (R&D) work on nanoparticulate metal oxide electrodes for fast-charging lithium ion batteries."

Okay, then. Good enough for the National Science Foundation is good enough for Goodenough — excuse me. I mean it's good enough for me. (Now excuse me while I take a victory lap for pulling off a topical setup/payoff pun for the ages. You're welcome.)

One dummy's verdict:

I give Battery Streak an 8-in-10 chance of coming to market in the next three years. I give it a 6-in-10 chance of coming to market in an EV in the next five years.

Battery Streak has three significant advantages going for it.

One is that the technology can be used on a multitude of platforms, like cell phones and the like. (Imagine taking your phone from 5 percent to 100 percent in five minutes.) The other is that it improves on existing lithium-ion technologies — you don't have to retool entire factories to capitalize on this technology. And the third is that it addresses perhaps the number-one fear of potential electric vehicle purchasers, which is the (perceived) time it takes to fill the tank with electrons.

Again, you and I both know now that this fear is because of an ICE fueling paradigm, and is therefore unwarranted, but then again people fear clowns despite very few clowns ever doing anything more sinister than making animal balloons and having leaky flowers attached to their lapels. Humans can be so vexing.

In any event, there are two places you should go to find out more and then simply keep any eye out for future developments:

- **The Battery Streak company website:** `https://batterystreak.com`
- **Matt Ferrel's YouTube overview of the same technology:** `www.youtube.com/watch?v=48vPgAPtkJg`

» Bracing yourself for cars that drive themselves

» Examining the current state of autonomous vehicles

» Managing competing visions for cars that can see

» Recognizing why self-driving is safer than you think

» Realizing that the future is now (and also just around the corner)

» Recognizing the challenges of our autonomous future

Chapter 11
Our Autonomous Future

Our future with autonomous vehicles is an exciting one to imagine. As you'll soon see, it's practically Utopian. In fact, let's introduce this chapter by articulating part of that vision.

Imagine this scene from the future:

You're checking your phone in the morning, shortly after you've walked the dog and had your first cups of coffee. After a check of weather and email, you pull up the Uber or Lyft app and request a ride to the office.

You then step outside just a few moments later as a vehicle (it'll almost certainly an EV) turns the corner of your block. You track the car's movement down your street, amazed at human creativity, all the way until the car stops at the at the foot of your driveway. And what makes this rather mundane-sounding scenario a vision of the future is what you see behind the wheel of your ride:

Nothing!

In our autonomous future, there's no human driving. In our autonomous future, all the cars can drive themselves.

And yet it's not after you hop in your ride that the wonders of autonomous driving really begin to manifest. After the car pulls away (after checking its surroundings, of course), you then get all that morning commute to do as you please. Want to finalize your presentation? Watch a movie? Play a game? Take a nap? With the knowledge that you're being monitored by the in-car cameras (something you've acknowledged in the app's terms-of-service), the time is all yours.

Even better: all this happens at a cost per mile that's much cheaper than what families typically pay to own a car. (I discussed the yearly cost for autos in Chapter 4.) For about 25 percent of what it costs to own an individual vehicle for a year, you can go anywhere at any time. Consider the implications of such a future. Imagine what you'd do with that extra $10,000 every year, or what that amount could mean to a family over the course of 20 or 30 years.

Better still: It'd be much safer than you driving to work. Your autonomous car would navigate roads full of other self-driving cars that would rarely get into crashes. (Current estimates place the chances for you getting into a crash on par with being hit by lightning.) Just *think* about never having to sort out an insurance claim ever again.

Beyond this, think about what autonomous vehicles might mean for those who have difficulty driving because of either law or disability — the blind, the aged, the young. It means never having to pick up your young kids from school at 3 P.M, or your grown kids from a bar at 2 A.M. It means virtually no more traffic congestion, because far fewer cars are needed to serve the community's transportation needs.

Beyond *that*, it means no more wasted land given to storing cars in parking lots. It means virtually no more parking lots — period. It means that a city like Los Angeles, California, suddenly has *three San Franciscos' worth* of real estate suddenly freed up for other uses — parks, housing, and more. Of course, if you don't need to park at the office, you don't need to park in your garage, either. You can take that space (and that savings) to significantly improve your home.

Oh, and I almost forgot about the revenue!

Say that you do decide to own a car that can drive itself. After it drops you off at work, you let *it* go to work. You have this glorious machine that can haul families to the airport, or drive older residents to the grocery store, or layabouts to the golf course (golf callback! I can't help myself), all while you're engrossed in your workday. Your own car shows up again at the office to cart you home. (Maybe it will tell you all about its day while you're in the back seat starting your happy hour just few minutes early.) By owning a self-driving car, you also own the potential for a lucrative side hustle.

If it works as imagined, it will be arguably the most transformative technological advance for humanity since — what? Iron? Agriculture? Vaccines? The automobile itself? It's quite difficult to overstate the impact that self-driving cars might have.

If it works as imagined. If.

Let's talk.

Adding Up the Self-Driving Benefits

If you're the type who skips over chapter intros, in this case, please don't. Not for this chapter. I've just spent a lot of space laying out most of the reasons that humans are hard at work trying to make this technology as much a part of our lives as indoor plumbing, refrigeration, or air conditioning.

Also, know from the outset that all the research and labor in the field of autonomous driving has already begun to bear fruit. As you will see momentarily, autonomous driving isn't a single technology that will one day get launched the way that the iPhone was in June of 2007.

Instead, self-driving — you may already have noticed that I'm using autonomous self-driving interchangeably — is the culmination of multiple interrelated technologies that are constantly improving. While the electric vehicle may be fairly analogous to an iPhone, a more accurate analogy for autonomous driving is medicine (or the medical field). That is, we've had medicine in the past, we have medicine today, and we'll have even better medicine in the future.

In any event, let's start with a discussion of the self-driving technologies that exist today (although "assistive" is probably a better word), and why we bother in the first place.

In one sentence? It's because self-driving is safer.

Or, phrased as a pejorative, it's because when it comes to driving, humans generally suck.

There are many reasons, so please spare some empathy for your fellow drivers, who probably want to get where they're going as quickly and as safely as you do.

For starters, driving requires a significant amount of one's cognitive functions, and those functions are negatively influenced by stress, fatigue, hunger, distractions, and even age. As I discuss in Chapter 6, your decisions behind the wheel at age 40 are about as good as they'll get, or at least that's what insurance company actuaries have concluded. At age 20 or 80, the insurance business says you're practically a menace when behind the wheel.

Self-driving technologies, on the other hand, don't ever get tired. They don't get fidgety when the lumbar support isn't set just right. They don't get drunk or take painkillers. They don't care if someone cuts them off in traffic or flips them the middle finger for driving too cautiously. They don't check their Instagram feeds while driving because they're not self-absorbed jerks. (My empathy well runs only so deep.)

Self-driving technologies don't even care if a kid is vomiting in the back seat, which almost caused yours truly to wreck many years ago. To a self-driving automobile, a kid throwing up is nothing that won't get sorted out by the parent in due time; the car will keep its unblinking camera-eye on the road, thank you very much.

The good news is that most cars sold today include many driver-assist features collectively known as ADAS, which is the acronym for Automated Driver Assistance Systems.

ADAS features found on today's cars include the following:

- Blindspot detection
- Front, rear, and cross-traffic alerts
- Collision avoidance
- Lane departure warning
- Forward and side collision warning
- Adaptive cruise control
- Park assist

Because these are available on today's cars, electric or not, and because the feature name itself is a pretty accurate description of what it does, I won't go into detail describing all features in the list.

And remember, autonomous vehicle technology is accretive — the first fully autonomous car will include features like lane departure warning, adaptive cruise control, and then augment those capabilities with tech that lets the car swerve to avoid road debris for example, or signal when changing lanes, or accelerate into the flow of traffic once the cross-traffic check has signaled that its safe (and legal) to proceed.

In short, the assistive-driving features available on almost all of today's EVs *already* deliver significant benefits. More on that next.

Listing the Key ADAS Benefits

For drivers, driver assistance technologies offer three key benefits:

- **Safety:** Data shows unequivocally that ADAS features reduce accidents. For example, about 30 percent of traffic collisions are rear-end crashes, according to the National Transportation Safety Board (NTSB). Many, if not most, of these, could be avoided or reduced in severity if all vehicles included automatic emergency braking systems. And that's just one safety feature, and one type of crash.
- **Convenience:** Having your car do a lot of the work for you reduces both fatigue and frustration. The car monitors a crowded highway with unwavering vigilance, sometimes applying the brakes for traffic conditions that have happened several cars ahead, and thus unseen to the human driver. Backup cameras significantly reduce the risk of a parking lot fender bender, which usually results in minor dents and major expense and headache.

  Beyond that, automatic parallel parking quickly feels like broadband Internet — experience it once and you'll never go back to doing it yourself.
- **Comfort:** Since the early days of cruise control, not having to keep your driving foot engaged while driving for long stretches makes the experience much more comfortable. Likewise, not having to grip the steering wheel can help cut back on muscle soreness, repetitive stress injuries, and even tension headaches.

REMEMBER

Although not a law on the books, but rather a gentlemen's agreement, it is significant nonetheless that 20 automakers, the National Highway Transportation Safety Administration (NHTSA), and the Insurance Institute for Highway Safety (IIHS) have pledged to make collision avoidance systems with automatic emergency braking standard on almost all new cars by September 2022. (Heavy-duty trucks have until September 2025.)

What's more, ADAS's safety benefits accrue to society at large. Fewer or less severe crashes — either with other cars or fellow humans — translates into fewer injuries, fewer medical expenses, and less loss of life.

So that's where we are today. This is the foundation which is being built upon.

Most of what follows, however, is looking out at what might be ahead. Most of what I discuss deals specifically with the topic of true self driving. As described in the chapter intro, true self-driving takes ADAS to entirely new levels of both safety and convenience.

Because if ADAS provides so much societal good just by *assisting* error-prone humans, how much more convenient and safer does driving become when we let the robots take over?

REMEMBER

If that last sentence sounded overly dramatic and/or dystopian, just remember that most of the past century has been about transitioning dangerous or repetitive work away from humans and to machines that do it faster and safer than we do. We used to have humans drive elevators. Now we've handed that task over to machines.

To answer that, we first have to deal with one overarching question.

Can These Cars Drive Themselves — or What?

The short answer is, "It's complicated."

As you read this, thousands of cars are most assuredly on the road — like right *now* — driving themselves without human intervention.

Some of these are from Waymo, which is a wholly owned subsidiary of Google. Waymo started making self-driving cars back in 2009, although retrofitting is more accurate — they converted Toyota Priuses. The company now has Chrysler Pacificas driving around Phoenix, Jaguar iPaces in San Francisco, and it has begun testing self-driving semitrucks in New Mexico and Texas.

All told, Waymo vehicles have logged over 20 million miles on public roads.

TECHNICAL STUFF

Is 20 million miles a lot? It's roughly the amount of miles that 2,000 people drive in a year.

Chevrolet's Cruise is another company that has deployed self-driving cars on city streets (San Francisco, specifically). Cruise and Waymo both use essentially the same technology on and around their cars, with the difference that Cruise deploys only Chevy Bolts that have been given names like Poppy, Burrito, and Pirouette.

Across the globe from where I call home, tech giant Baidu has begun deploying Apollo driverless taxis (robotaxis), making it the first company to have a commercialized autonomous driving service in China.

And Domino's Pizza has started driverless pizza deliveries in places like Houston using a vehicle called a Nuro, pictured in Figure 11-1.

FIGURE 11-1: Hot and fresh and straight to your — "OMG, no one is driving that thing!!"

Dominos pizza

Other self-driving vehicles, as you may be aware, are Teslas. For better or worse, Tesla has taken the self-driving concept from a rather opaque engineering pursuit and thrust it into the global spotlight through a combination of its CEO's bold promises, bold pricing, bold technological choices, and some decidedly hit-or-miss execution.

To be sure, Teslas can drive from Point A to Point B without (much) human interaction. When equipped with the optional Full Self-Driving feature, they can pull out of tight parking spots with no one behind the wheel, something my family has done several times when navigating the notoriously tight parking spaces of San Fran. That feature in particular is called Summon, and I can tell you from experience that it feels like you're commanding a giant radio-controlled (RC) car, except that your phone is the controller.

So, yes, in short, cars can drive themselves today. However, all three of the self-driving cars I mention here still require a human behind the wheel to supervise proceedings. (I guess the Summon feature is an exception.) In other words, full self-driving isn't quite yet to the level of *full* full self-driving.

Interestingly, self-driving is a matter of degree. Or, more accurately, a matter of *level*. And the good news on this front is that any electric vehicle or hybrid sold today includes at least *some* level of self-driving. (Here's where all that ADAS foundation will come in handy, BTW.)

Since that foundation is in place, we're ready to talk about what those levels are.

Leveling up your self-driving knowledge

As the old saying goes, there are two types of people in this world — the kind who puts everyone into two groups, and everyone else. This section is dedicated to the everyone else among you who are about to learn about six different levels of autonomous driving.

These six levels have been defined by the Society of Automotive Engineers (SAE), and, as we all know, engineers just *lo-o-ove* their categorization systems. But apparently it works, because these categories have been adopted by the US Department of Transportation as well.

Categories range from 0 (fully manual) to 5 (fully autonomous). The first three levels are available on almost any new car, EV or not.

Level 0: No Driving Automation

With Level 0, the human does all the driving, but there are systems in place that assist. These include warnings for blind spots, lane departure, and automatic emergency braking (AEB).

AEB is considered Level 0 because it doesn't "drive" the vehicle — human hands are on the wheel and feet are always on the pedals.

Level 1: Driver Assistance

This is the lowest level of automation (and explains why the SAE starts this scale with Level 0). Level 1 features a single automated system for driver assistance, like adaptive cruise control and/or lane centering.

Even when one of these systems is active, the human monitors other aspects of driving, like braking and steering.

Level 2: Partial Driving Automation

When automobile manufacturers refer to ADAS, this level of automation is usually what they mean, Level 2 automation means that the car can control both steering and speed simultaneously.

Some Level 2 systems allow you to drive entirely hands- and feet-free, although a human must sit in the driver's seat, ready to take control at any time.

Level 2 Examples

We pause here to note some of the current Level 2 systems, which include Ford's BlueCruise (a name that reminds me of a kid's TV show whose host(s) wore green rugby shirts), GM's Super Cruise (which reminds me of a vacation to faraway ports of call), and Tesla's Autopilot (which reminds me of a *30 Rock* episode featuring Matt Damon).

Autopilot can even change lanes for you when using the turn signal; expect BlueCruise and Super Cruise to follow suit.

TIP

It's totally worth ten minutes of your day to look up the backstory of *Blue's Clues*, by the way. If not, are you really using your Internet to its full potential? I didn't think so.

How do Level 2 systems ensure that "a human must sit in the driver's seat, ready to take control at any time?" The answer is, with a camera pointed toward the cabin, making sure you're paying attention to the road.

Let me repeat that statement: When using Level 2 autonomy, *the car is monitoring/filming you at all times.* It can even track your eyes when you're wearing sunglasses.

And if the personal privacy activist in you can't quite stomach the idea of being watched at every moment while driving, let me gently suggest that you should forego your smartphone, which tracks your every move. You also might want to unplug your smart speaker, which listens to every one of your conversations. (But hey, it makes it easy to check the weather.) Oh, and never do a Google search or use the Internet without a VPN or post something to Facebook, although that won't really matter — Facebook tracks most of your web activity anyway, even if you delete your Facebook account, which by all means you should. What I'm saying is that there's a trade-off with all these sorts of technologies. It's cool to be able to learn about Steve, Joe, and Josh from *Blue's Clues* 15 seconds after I kinda *Inception*-ed that into your brain. It's also cool to let a car drive itself down the road. What's not cool, yet unfortunately not beyond some people's capacity for stupid, is to do have the car start driving itself given the current state of the tech, and then jump into the back seat to watch *Blue's Clues* reruns. Hence the camera.

The next three levels of autonomy move the conversation from the realm of partial autonomy into the utopian vision of full autonomy. As should come as no surprise by now, the conversation is one of aggregation, where the next level of autonomy adds features to the previous one.

Level 3: Conditional Driving Automation

Level 3 automation is kind of tricky to describe, and the difference is almost negligible to humans when weighing it against Level 2.

A Level 3 car has awareness of environmental conditions, enabling it do things like pass a slow-moving vehicle or navigate around a delivery truck that is stopped in the road. Level 3 autonomy still requires humans to take over, though, if the system is unable to execute the task.

Are any cars today capable of Level 3 autonomy? Yes. In late 2021, Mercedes-Benz became the first company whose Drive Pilot self-driving system was granted the UN-R157 designation. The UN part of that designation is for the United Nations, and the R157 means Level 3 autonomy, I suppose.

In any event, the Mercedes Level 3 system is available on the S Class and the all-electric (yay!) EQS. (See Figure 11-2.)

Level 4: High Driving Automation

Now we're getting close — you can see Utopia on the horizon from here. With Level 4 automation, the car drives itself in *most* conditions, although the human is still there to manually override when the computer can't quite figure things out.

One of the biggest obstacles to Level 4 driving is the legislation involved in unleashing onto public roads cars that don't require a driver. So far, they've been *geofenced* (i.e. having a virtual geographic boundary, in other words), and are intended for ride-sharing.

What about Level 4 today? Can I get one of those? Yes and no. Yes, you can hail a Waymo, which is indeed Level 4, in a preapproved area like portions of Phoenix,

New York, or San Fran. No, you can't tell your Waymo to drive you to Vegas while you're in the backseat reading Kerouac. And also, no, you can't buy a Waymo for personal use.

FIGURE 11-2: EQS drivers use Level 3 autonomy. S-Class drivers could use Level 3, but probably would prefer a chauffeur.

Mike Mareen / Adobe Stock

What's more, several other companies — Navya, Magna, Lyft, Volvo, and Baidu — are building/selling/partnering on Level 4 vehicles and shuttles (hopefully in Vegas, where these things are *really* needed).

Watch this space. As I said, we're getting close.

Level 5: Full Driving Automation

A Level 5 autonomous vehicle drives itself — full stop.

In fact, *vehicles* is much more accurate than *cars* in describing Level 5, because Level 5 cars may not even have a pedal or a steering wheel. A Level 5 vehicle can drive anywhere, generally replicating the behaviors of an experienced human driver.

So Level 5 is the holy grail humans pursue. The Nirvana. The Utopia. Level 5 autonomy unlocks all the possibilities I mention in the intro to this chapter: more money, space, and time for humanity. Level 5 autonomy enables Transportation as a Service (TaaS).

It's an extraordinarily complex problem. And it's being tested around the world by really smart, really dedicated men and women. I think the odds are very much in humanity's favor that we will eventually succeed in the task. I think the computing power is there. I think the code is getting close. Fifty years ago, people predicted that computers would never be able to play chess better than humans. Or Go. Both predictions were wrong.

And, in another 50 years — assuming that we have a habitable planet by then — we won't be able to remember a time when we had to actually operate a car with our feet, eyes, and hands. The trajectory of technology tells us that humans eventually figure out how to make machines better at repetitive, dangerous tasks.

But no. As this book goes to press, nothing out there can be considered Level 5, and I discuss some of the fascinating reasons why in the sections that follow.

Doing the Vision Thing

Carmakers use different approaches to give their cars a form of sight — to do through sensors and processing what humans do with their eyes and brain. And an active argument is taking place right now about the best way to do this.

This topic is important because the car you buy today may get upgraded software in the future, but it's unlikely to get different sensors and different processing capabilities. In a certain sense, what the car "sees," on the day you receive it, is what you get, in terms of the future sensing capabilities of the car. And that may enable, or limit, how far it can climb the vehicle autonomy ladder I describe in the preceding sections.

So here I briefly describe the technological approaches used by several major players so that you can understand some of the discussions going on today, and likely to continue for years to come — and then go make the right purchasing decision for yourself.

Tesla and Comma.ai bet on cameras

The human eye includes neither radar nor lidar. And neither do newer Teslas.

Radar sends out electromagnetic pulses and then measures their return to a sensor to judge the distance, solidity, and velocity (if any) of outside objects. Lidar, short for *li*ght *d*etection *a*nd *r*anging, does the same thing, but with laser beams.

With both radar and lidar, the bigger the pulse emitter, and the bigger the return sensor, the better the results.

For more than a decade, from Tesla's launch in 2008 until 2020, it used small radar sensors, ultrasound sensors, and cameras in its cars. These cars supported a range of ADAS features, including automatic emergency braking, forward collision warning, lane departure warnings and avoidance, and blind spot/side collision warnings. But in 2021, for the North American market, Tesla switched to Tesla Vision.

Tesla Vision is based entirely on cameras and ultrasound sensors — no radar and no lidar. In fact, the company has come out against both, taking the position that humans don't need radar or lidar to see, so why should cars? (Humans also don't use ultrasound, except perhaps for Batman, but that doesn't stop Tesla from using ultrasound sensors in its cars.)

The initial implementation for Tesla Vision has 8 surround cameras and 12 ultrasonic sensors. It's used only in the North American market at present. Tesla faced challenges with the rollout of Tesla Vision, because it took time for regulators to fully approve the forward collision warning and automatic emergency braking features on cars with the new set of sensors.

Earnest debate takes place about how far a no-radar approach can go in delivering higher levels of self-driving capability. What is not in question is that cameras and ultrasound sensors are small and cheap, compared to radar. What is also not in question is that Tesla has more electric vehicles on the road, driving many more miles, than any other vendor. If anyone can make advanced self-driving features work entirely with cameras and ultrasound sensors, it should be Tesla — but I am frankly skeptical.

comma.ai is a far different company that is taking the same general approach as Tesla. comma.ai releases its product as open source software that runs on standard smartphones. Smartphones do, after all, have multiple cameras, a bit of infrared, and a minor amount of lidar. But I am really skeptical that a system built around a smartphone can really do the job.

A wide range of recent cars can be driven "by wire," and comma.ai works with those cars, using a connector called panda, plus the sensors in your phone. (Driving something *by wire* means that there are no mechanical controls. A wire sends a signal to the physical components that actually perform the braking, accelerating, and steering. Airplanes, for example, are now flown by wire. The yoke doesn't directly control the wing flaps; instead, turning the yoke sends a signal to the machinery that does.)

SELF-DRIVING AND THE THREE LAWS OF ROBOTICS

Long before humans started developing cars that could drive themselves, science fiction writers were contemplating the ethical implications of robots being imbued with decision-making capabilities. This speculation was most famously framed by Isaac Asimov in his seminal novella *I, Robot,* which introduced the world to the Three Laws of Robotics:

- A robot may not injure a human being, or through inaction allow a human being to come to harm.
- A robot must obey orders given by human beings, except where such orders would conflict with the first law.
- A robot must protect its own existence as long as such protection does not conflict with the first or second laws.

A fairly logical framework, yes? Will probably only take a couple lines of code.

Except: Suppose that a self-driving car is cruising down a typical suburban arterial doing 45 mph when a child suddenly runs across the street. The only way to avoid a collision with the child is to swerve into the oncoming lane of traffic.

How does a robot behave in this situation? (And, in this macabre hypothetical, bear in mind that humanity has achieved full Level 5 autonomy. We also assume only one adult passenger in the vehicle.)

Option 1, hit the brakes but continue on the current course, which results in the child being struck, which violates Law 1.

Option 2, swerve into oncoming traffic, which violate Laws 2 and 3.

If it's any consolation, know that programming a concept like "harm" into a computer is, well, impossible. If the car hits the child and spares both itself and the passenger, what is the psychological "harm" to the car's passenger?

And what if, instead of a child running into the street, it's a bank robber, gun drawn, running toward a woman in a stopped car? Now what does the car do? Computers are good with math problems like compound interest. They're still terrible at recognizing nuanced human concepts like harm or justice.

GEORGE HOTZ, COMMA.AI BOY WONDER

If you're the type who likes reading up about young, socially awkward supergenius types like Elon Musk, then boy-o-boy, do I have the dude for you to spend a few hours with on YouTube. At the time of this writing, George Hotz is a 32-year old "American security hacker, entrepreneur, and software engineer . . . known for developing iOS jailbreaks, reverse-engineering the PlayStation 3, and for the subsequent lawsuit brought against him by Sony."

In 2007, 17-year old geohot — one of his several online handles — became the first to remove the SIM lock on an iPhone. He then traded his second unlocked 8GB iPhone to Terry Daidone, the founder of CertiCell, for a Nissan 350Z (not an electric car).

Oh, and he founded comma.ai. So, yeah.

George looks/dresses like a Hollywood actor playing a computer hacker — sort of a Scott Lang to Elon Musk's Tony Stark — and he speaks like he's visiting Earth from another plane of existence, kind of waiting impatiently for six dimensional words to downrez to their 3D counterparts.

Perhaps most significantly, he has posted 28 or so hip-hop songs to SoundCloud under one of these other handles, tomcr00se. Because of course he did.

comma.ai users are logging hundreds of thousands of miles using the system, and the company is analyzing video from these volunteers to gradually improve its system.

Waymo and Cruise add lidar

Waymo started as Google's self-driving car program, becoming a separate company in 2009. Its cars are familiar on the streets of Phoenix, San Francisco, and New York. (See Figure 11-3.)

A Waymo car has a cone on top that looks like an upside-down bucket. Inside the bucket is a lidar unit. The bucket is highly capable; it can be tuned to various frequencies that allow the system to "see" at closer or greater distances. The input from lidar is fed into the same system that processes all camera and ultrasound sensor input.

HOT TAKE: THIS IS ALL VERY STUPID AND WILL NEVER WORK

I can see a future where self-driving gets people and products from Point A to Point B along major highways safely, serving a similar function as railroad lines do today, and likely with a similar degree of regulation.

However, I can't quite wrap my mind around how self-driving cars will solve what I call the Asimov problem, which is that they'll be programmed to avoid collisions with humans at all costs.

So then, in a city environment where cars aren't moving all that fast to begin with, why would anybody walking ever yield the right-of-way to a car? In a downtown environment, why not just cross the street regardless of traffic, knowing full well that cars will slam on the brakes rather than hit a pedestrian? (In fairness, I guess humans would try not to hit pedestrians, too. In theory.) But in a busy, walkable city where no one yields to cars, how do the cars do the job they were designed for?

FIGURE 11-3: Either an old Waymo or a newer Dominos Nuro — difficult to tell them apart.

Tada Images/Adobe Stock

So Waymo is throwing a lot of hardware at the problem, and the company is seen as the industry leader in self-driving capabilities. It's the only company running driverless cars on the road at any kind of scale, in its robotaxi program, almost entirely confined to the cities just listed.

Cruise is one of several other companies also using advanced hardware and running a robotaxi program to deploy it. Neither company is making cars for consumers, as Tesla is.

Other carmakers split the difference

Major carmakers are using cameras plus small, short-range lidar sensors, provided by a range of start-ups, in systems such as Ford's BlueCruise and GM's Super Cruise. These efforts are still on few enough cars that the technology is not yet in full production, and no one can be sure what the cost and performance parameters will evolve to be in the years ahead.

Tesla and comma.ai are the most aggressive in pointing toward full self-driving capability, and Tesla claims that the car you buy today will be able to be upgraded to full self-driving via software update, which is inherently much cheaper to provide than a major hardware upgrade could be. comma.ai says you'll be able to do the same thing with your smartphone.

Waymo and its close competitors are seen as today's leaders in self-driving capability. This sounds promising, but these companies are not in the consumer EV market. Perhaps they hope to sell to consumers in the future, but if so, it will be many years before they can be in production at volume.

Other carmakers, which use lots of cameras plus small, low-end sensors, are delivering Level 2 capability today. They are not making aggressive promises that their cars will be able to go beyond that point, using the hardware that comes with the car you buy now.

So the car you buy today has to deliver the capabilities the carmakers claim to have today, or else they would get severe pushback from the marketplace and regulators. You can definitely make a solid buying decision based on that.

What's at issue is whether you should also let future promises of advanced capabilities, like those made by Tesla, influence your buying decision. I can only leave that up to you, but promises about tomorrow may not be a solid basis for choosing where to spend your money today.

Recognizing the Challenges

None of this will be easy.

REMEMBER

Never mind the cars and the cameras and the software. None of this will be easy from a data throughput standpoint. According to analysts at Morgan Stanley, autonomous vehicles will transmit massive amounts of data through cellular networks, meaning that AV EVs *must* have access to wireless service 100 percent of the time, something which is not currently possible.

For further reading, check out: `www.bloomberg.com/news/articles/2021-09-17/carmakers-look-to-satellites-for-future-of-self-driving-vehicles`.

EVs themselves represent amazing advances in technology, and the world is still struggling with providing the charging stations needed to power them, even as the need for lower-pollution transportation becomes clearer by the day.

Self-driving capabilities are a whole 'nother level of challenge. Autonomous cars have problems, and so do those regulating them. Let's take a moment to unpack each in sequence.

What's still really hard for autonomous cars

Let's pretend that we're all software developers. What does a software developer do? It tells computers what to do. Rudimentary, but it'll do for our purposes here.

With that in mind, let's tell a self-driving vehicle what's it's seeing in Figure 11-4:

What do we tell a computer program this is, though? A computer doesn't know what a fire hydrant is. Computers have to be *told* what a fire hydrant is in terms like these:

- 36 inches tall
- Red
- Conical top

Great! We've just programmed a computer to recognize what a fire hydrant is. Let's now go to the next item in the backlog item, conveniently displayed in Figure 11-5.

This, of course, is a small child bundled up for a day of sledding or eating snow, but we've got the hang of it now. How do we describe this small child to a computer program? Easy, it is:

» 36 inches tall

» Red

» Conical top

FIGURE 11-4: A fire hydrant.

Andrew Ferguson / Adobe Stock

FIGURE 11-5: A fire hydrant — wait, no, this is a child! This is so complicated! How do computers figure out the difference!? (Actually, that was a trick question. Computers don't figure anything out. They just carry out instructions.)

Andriy Medvediuk / Adobe Stock

Wait.

That's the same list. So how is the computer supposed to know the difference between . . . oh, right. This is why programming a computer to a) recognize and then b) make *decisions* about what it recognizes is such a difficult, and literally life and death, problem to solve for autonomous driving.

(To the AI experts out there: I'm fully aware of flaws in the analogy I've just laid out; I'm just trying to establish a baseline that this is all really, really problematic for conventional computer programming.)

Cars with ADAS features — Levels 0, 1, and 2 in the earlier self-driving roadmap — have trouble with such surprises, including pedestrians who appear from out of nowhere, emergency vehicles blocking part of a road, concrete barriers around a closed lane on a highway, or distinguishing between fire hydrants and overdressed toddlers.

For cars to move into semi-autonomy and autonomy — Levels 3, 4, and 5 in the self-driving roadmap — these problems need to be solved in the automated part of the system because with these self-driving levels, the human is no longer in the loop.

The bottom line is that "better than human" capability doesn't seem to be good enough for the public. "Nearly perfect," is the seeming bar to clear for semi-autonomy and full autonomy to become reality. And perfection, or even near-perfection, is famously difficult to achieve.

What's still really hard for humans regulating autonomous cars

Regulators for the auto industry and public roadways have it really tough. In its early years, Facebook's motto was "Move fast and break things." This motto pretty much sums up what regulators don't do, and it captures a lot of their difficulties in regulating more or less autonomous cars.

Regulators don't usually move fast. They are under tremendous pressure to not make mistakes. Studying a new technology more, consulting more widely, and waiting for the next release all reduce the chance of making a mistake. So regulators are often unmoved by the need of a carmaker, for instance, to get a car to market by year's end.

And regulators are specifically tasked with not breaking things, and especially not allowing large numbers of accidents — events that damage people, not just the cars they're driving in. It's the regulators' job not to allow things to be broken.

Anyone familiar with the way regulation works might well be amazed that Tesla can make the claims it does, or that Waymo can run self-driving taxis, given the very real risks involved. Regulators are allowing a lot of innovation to happen relatively quickly. If pushed too hard, they may be more inclined to slow down the current pace of change than to speed things up.

Gazing into the EV Crystal Ball

Progress on EVs is not as fast as the environment needs it to be — but, with Tesla in the lead, it has been quite rapid, and may accelerate further. Autonomous vehicle capability has also moved quite quickly.

Where AV capability is coming fast

A car that can change lanes by itself seems pretty darned advanced, from my point of view. And major carmakers are right on the verge of selling such cars worldwide.

Strong safety improvements have been achieved for Level 0 features such as automatic braking, and several Level 0 features are now required in most markets.

If safety improvements continue with Level 1 and Level 2 features, expect them to also quickly move from extra-cost options to required features.

The inflection point from Level 2 and below to Level 3 and above seems to be a more difficult puzzle. But again, when a feature reaches a point such that it demonstrates strong improvements in safety, expect it to gain widespread adoption rather quickly.

Where AV has already arrived

AV capability is widely available in two forms today.

Every Tesla can be upgraded to Level 2 self-driving, including the ability to change lanes on its own as directed by the navigation system.

From the point of view of a decade or two ago, that is an almost shocking level of capability. And Ford, GM, and other carmakers seem to be on track to match this feat in some of their newer models. A car with Level 2 ADAS features will be much safer, more convenient, and more comfortable to drive than the 100-plus years' worth of cars that preceded them.

The challenge they collectively have is to solve is moving to Level 3 and beyond, though full Level 2 capability is a considerable safety and convenience accomplishment on its own.

At the higher levels, Waymo cars are driving themselves on the streets of a few cities today, giving customers rides and making deliveries. This is truly amazing when you think about it. The challenge that the Waymos and the Cruises of the world have is whether this advanced technology can scale outward, across more cities and entire countries, while also moving down-market — into vehicles that are sold to consumers.

While these visible achievements are ongoing, big companies and hungry startups are working hard behind the scenes. They're striving to improve — and to drive down the price of — every single hardware and software component needed to make AV capabilities affordable. They're integrating these pieces into turnkey systems (in the same way that software maker Adobe has a turnkey solution for desktop publishing or video editing) that EV makers can easily drop into their cars.

So. Are we ready to implement the utopian vision of fully self-driving cars? Are we ready to nap in the backseat while on the morning commute, and to make personal transportation much less of a financial burden, and reuse public right-of-way that's otherwise given (wasted) for vehicle parking?

Not quite yet.

Will we get there eventually? I think the odds are in our favor. I eagerly anticipate our autonomous future.

» Looking at the pluses (and minuses) of electric buses

» Going the distance with electric semis

» Adding electric delivery vans to the mix

» Dumping the gas-guzzling lawn mower

Chapter 12
Other Vehicles That Are Going Electric

It should be obvious by now that I'm in league with those who argue that electric cars are the future of personal transportation.

But you may also notice that cars aren't the only wheeled vehicles you'll see along roadways. Some of the other vehicles have four wheels, just like cars do. Others have six. Some have eighteen.

So, what other kinds of electric vehicles might you see traversing roadways sometime during this decade?

Let's take a look.

Going Mobile with Electric Buses

From an engineering standpoint, electric buses behave the same way that electric cars do — an EV mass transit bus is essentially a longer, taller version of the Volkswagen ID. Buzz.

Like electric cars, these buses mostly consist of a battery, some electric motors, and a steering wheel. When you put all these items together, you end up with a people mover that is quieter and offers better performance than a diesel-powered bus. In fact, electric-bus passengers emit more emissions during a trip than the bus does, which is kinda gross, I guess, but bodily functions are bodily functions — I'm not one to judge.

In all seriousness: You don't exactly have to study air quality for a living to notice the black soot coming out of the exhaust pipe of a diesel bus and think, "Hey, that's probably not the best thing in the world to breathe." If you live in a metropolis plagued by air quality issues (see Figure 12-1), all you have to do on some days is step outside and the case for electric buses — electric everything, really — is literally in the air.

FIGURE 12-1: Switch to electric buses? And miss out on lungs full of *this?*

kichigin19 / Adobe Stock

Until now, the main obstacle to electric bus adoption has been a balancing act between battery pack size and weight. If the battery pack is too small, the bus can't make it back-and-forth on its route. But the problem can't quite be solved by "more batteries." At a certain point, it doesn't make much sense from a financial standpoint to keep adding more batteries, and at a certain point additional batteries make the bus too heavy, and heavy vehicles put more wear-and-tear on the road.

But the same advances in lithium-ion batteries that have enabled their use in passenger vehicles have been applied to *b*attery-*e*lectric *b*uses (BEBs), leading to the reductions in weight and increases in range that are necessary to make electric buses a viable solution.

Which is why BEBs are already on city streets. For the most part, they look like any other buses in service. (See Figure 12-2.) And, if your city's roadways are properly managed, this bus is also the *fastest* way to travel around a city or town.

FIGURE 12-2: The wheels on the bus go round and round.

AB27 / Pixabay

At the time of this writing, China operates over 90 percent of the world's battery-powered buses, although that percentage is quickly shifting as other municipalities adopt EV buses.

The city of Detroit, Michigan, added four electric buses to its fleet while I was drafting this chapter, for example. Indianapolis, Indiana is likewise busy deploying a fleet of electrics made by BYD (which is discussed later in the chapter). And I've recently returned from a short stay in Banff National Park in Alberta, Canada, where I discovered a few amazing hikes amongst the Canadian Rockies, and some electric buses that could quickly get me to the trailhead.

What's more, the EPA has recently opened submissions for its Clean School Bus Program, which will supply funding to school districts throughout the US if the districts replace old diesel buses with electric or liquid natural-gas-powered alternatives. Schools can earn $375,000 if they make the switch to electric.

The bottom line to all of this is all about the bottom line: Electric buses are taking hold because they're much cheaper to fuel, which makes them fantastic.

TIP

If a car generally gets driven 40 miles or so in a day, then a bus must be driven for — what? — 500 miles? A thousand? How many miles does a bus cover in a day?

The answer is somewhere between 150 and 250. I was surprised to learn that the number was that low, but it doesn't mean that BEBs with current battery technology can travel that far in a day. A common battery pack size is 240 kWh, which yields, driving with an efficiency of 2 kWh per mile, 120 miles of range.

More about that topic in the next section.

Seeing why battery-electric buses are fantastic

Decisions about bus purchases, whether diesel or electric, are made not at the individual level, but rather at the municipal level. The city councils and metropolitan transit authorities involved must therefore weigh several factors when buying machines used to transport city residents and guests from place to place.

One of those considerations is a line item on every transit authority's balance sheet: the cost of fuel. This is where electric buses offer a significant value proposition to budget directors. As I point out throughout this book, electric batteries are much cheaper to fill than gasoline tanks. Multiply this by the fact that most buses are in service seven days a week, and then multiply again by the number of buses in service — the per year savings can potentially run into the seven- or eight-figure range.

Another consideration you might be thinking of turns out to be not much of a consideration for municipalities operating bus fleets: time to recharge the battery.

Yes, charging a giant 240 kWh battery can take between 2½ to 4 hours, but cities solve for that length of time by simply buying two buses for a single route. One bus can charge and watch Netflix, or whatever buses do when they're not being used, while the other EV bus is driving along its route. Also, some routes/buses, known as *trippers* to people who do this stuff for a living, are in service 3 or 4 hours during morning rush and 3 or 4 hours in the evening. In this scenario, a single EV bus can charge up again midday and overnight, beginning each of its trips with a full battery.

Finally, there's an argument to be made for the use of fast charging to keep electric buses continually in service. I haven't yet been able to find examples of it out in the real world just yet, but the logic of such a system is sound. It goes like this:

- Most transit buses travel 11 to 13 miles in 1 hour.
- The buses repeat their route every hour or so.
- The buses return to a common point and have a 5- to 10-minute layover.

A bus can then simply fast-charge during the layover, adding an hour's worth of energy in about 5 minutes while passengers are exiting and boarding. Even if the bus were driven with an efficiency of 1 kWh per mile, a 5-minute fast charging session should allow plenty of time for 13 kWh of energy to be put back into the battery. (A 250 kWh DC fast charger can add 250 miles per hour, or roughly 4 kWh per minute. Five minutes at this speed can put 20 kWh back into the battery, which would be plenty for this typical use case.)

Another use case that might be able to leverage fast charging is airport shuttles. The next time you're at the airport, notice the number of shuttles ferrying travelers to and from parking lots or rental car facilities or even between terminals. Most of these shuttles are smaller than a typical city bus, and the shuttles usually make trips of only a few miles or less. In my mind, this is the perfect use case for a lightweight electric bus/shuttle fleet. I'm sure that accountants at Hertz, Dollar, Avis, and other businesses are shopping around and making calculations as I speak.

However, as with so many technologies of the future, there's always a *but.*

In this case, there are several.

But, but — electric buses aren't all that fantastic, after all

The first *but* is that electric buses don't necessarily *have* to draw from a battery the power they need to make them move. Instead, they can draw power from the electric grid itself.

This kind of arrangement is called a trolley bus, and trolley buses are traversing all over many downtowns all over the world, from San Francisco to Oslo to Sao Paulo, Brazil. And they have been doing so for over 100 years. (See Figure 12-3.) The first trolley bus debuted in Berlin in 1882.

FIGURE 12-3: The future of electric buses is . . . 150 years old?

The important thing to remember is that trolley buses *are* electric buses. They have no gasoline engine, but neither do they have a battery pack. They acquire the energy they need to run their electric motors by connecting two trolley poles to two electrical wires overhead.

As you can see, trolley buses are best used when the route is heavily trafficked and doesn't change for 99 to 100 percent of the bus's operating life. Picture a bus that travels back-and-forth along a single street or two, as is the case with the many routes that go up and down San Francisco's Market Street.

One significant advantage of electric trolley buses is that they're much lighter vehicles than BEBs, and lighter vehicles are much more forgiving of road surfaces. After all, savings from the cost of fueling can be negated by extra expenditures for road maintenance.

The other advantage of a trolley bus is that it never has to be taken out of service to charge the battery. As long as the tires are inflated and the wires are connected, the trolley bus is good to go.

When zooming out to the overall electric transportation landscape, you can see that different use cases have different solutions from using current technology. A battery-powered bus is probably not ideal for a dense downtown thoroughfare.

However, for vehicles like school buses, or for trips between Banff Springs and the Banff Gondola at Sulphur Mountain, the only workable solution is to start making the switch wherever possible and continue to improve battery performance.

Another type of electric vehicle that will be entirely dependent on batteries continuing to get cheaper, lighter, and more energy dense is the long-haul truck. More on them in the next section.

Keep On Truckin' (with Electric Long-Haul Trucks)

Electric semis — also known as electric long-haul trucks, also known as electric tractor trailers, also known as electric 18-wheelers, also known as electric big rigs — will soon get their day in the headlines, for reasons other than some announcement of delayed delivery dates or deceptive claims about how close they are to being mass-produced.

And, if that last sentence leaves you a bit confused because you don't stay abreast of electric trucking news the way yours truly does, stay tuned.

TECHNICAL STUFF

Yes, 18-wheelers actually have 18 wheels — two front wheels used to steer the rig, and then 8 on the dual axles of the cab which also support the trailer's front weight, and then 8 more to support the trailer's rear. Electric semis will have their well-deserved day in the sun for good reason. Because, although a passenger car is in operation for only an hour or two each day, a long-haul semi can, and often is, driven for the maximum of 11 hours per shift.

As with electric buses, it all comes down to the cost of fuel. A typical diesel semi gets between 6 and 8 miles per gallon, or mpg. When multiplied by the cost of diesel, which is about $7 per gallon in California as I draft this chapter, there's an urgent market incentive to convert gasoline trucking over to electric.

Because semis are on the road for many more hours per day than cars are, replacing one semi with an electric version thereof has the environmental and financial impact of a dozen or more electric car sales.

But once again, these things are not yet ready for a starring role in a retconned *BJ and the Bear*. And once again, it's all about the battery. (And also a little bit about autonomous driving.)

As this book goes to press, the battery technology hasn't quite delivered on the range and charging capabilities that would make long-haul electric trucking a viable replacement for current tech. As it stands, fully loaded semitrailers simply require too much energy to start them moving and keep them moving for the amount of range that most long-haul trucks cover between stops. For most long-haul applications, you need to somehow harness the power of dynamite, which gasoline does, as you may have read earlier in this book.

But there are fortunes to be made for those who can address the problem. I describe in the sections that follow some of the main firms competing for a slice of that money pie.

Tesla Semi

Year of debut: 2017

Expected delivery: 2023

Stated range: 300 to 500 miles, but I'll believe 500 when I see it

Major customers: Anheuser-Busch, Pepsi, UPS, FedEx

Unlike most vehicles, never mind most trucks, the Tesla Semi (see Figure 12-4) places the driver in the center of the cab. Tesla also announced the Megacharger, which would be used to charge the Tesla semi's large battery packs — theoretically, in a similar time frame used for a typical stop.

FIGURE 12-4: The Tesla Semi.

Mike Mareen / Adobe Stock

Freightliner eCascadia

Year of debut: 2018

Expected delivery: Late 2022 (so 2023, at the earliest)

Stated range: 250 miles

Major customers: Penske Truck Leasing, NFI Industries

The Freightliner eCascadia (see Figure 12-5) is manufactured by Daimler Trucks North America. The semi is based on Freightliner's Cascadia diesel long-haul semi, and its primary use case is for deliveries to and from a regional hub, where it can then be charged overnight to full capacity.

FIGURE 12-5: The Freightliner eCascadia.

Volvo VNR Electric

Year of debut: In its current iteration, 2022

Expected delivery: 2023, but it's anyone's guess

Stated range: Up to 275 miles

Major customers: Unknown

Like the Freightliner, the Volvo VNR (see Figure 12-6) isn't exactly engineered for long-haul trips — say, a route across the continental United States. Instead, it's meant for shorter, more predictable routes from a central distribution hub, where it can park and recharge overnight (or at least at the end of every shift).

FIGURE 12-6: The Volvo VNR Electric.

Jan Kliment / Adobe Stock

Kenworth T680

Year of debut: In its current iteration, 2022

Expected delivery: Not announced

Stated range: Up to 150 miles

Major customers: Unknown

I made an esoteric *BJ and the Bear* reference earlier. I don't know what kind of semitruck BJ actually used, but I do know that Kenworth semis were used by Burt Reynolds in *Smokey and the Bandit*. And I know that Kenworth will do its best not to be sidelined when it comes to the BEV trucking business.

No one is engineering the Kenworth T680 for long-haul freight routes, but rather for regional or depot kinds of application. (Are you sensing a theme for these BEV semis yet?) For a peek at the T680, check out Figure 12-7.

BYD 8TT

Year of debut: 2021

Expected delivery: In operation now

Stated range: Up to 167 miles

Major customers: Unknown

FIGURE 12-7: The Kenworth T680.

The BYD semi (see Figure 12-8) is in operation today, serving as a workhorse for regional freight hauling, local distribution, and port-to-terminal routes. What's also interesting about the 8TT is that it uses iron phosphate batteries rather than lithium-ion because iron phosphate is one of those future/now battery technologies, as discussed in Chapter 10.

FIGURE 12-8: The BYD 8TT.

As motioned earlier in this chapter, transportation authorities in Indianapolis, Indiana are very familiar with this company.

And, as I discuss in Chapter 15 (it's a Part of Ten), the Chinese automaker BYD is one of the largest electric-vehicle manufacturers in the world. Besides electric semis, BYD makes electric garbage trucks, buses, forklifts, and more.

Nikola Tre

Year of debut: 2017

Expected delivery: December 2021

Stated range: Up to 350 miles

Major customers: Anheuser-Busch, Total Transportation Services

As I'll touch on, Nikola is a company that has been mired in controversy yet has seemingly navigated the rough waters churned up by its founder, Trevor Milton — waters here serving as euphemism for "paying a $125 million fine in order to settle fraud charges from the Securities and Exchange Commission."

In any event, Nikola's Tre is a Class 8 semi that travels farther on a single charge than do most other competitive truck models. (See Figure 12-9.) The company began pilot deliveries of the Tre in December of 2021, and production continues at its Ulm, Germany, manufacturing facility.

FIGURE 12-9: The Nikola Tre.

Stefan Puchner / AGE Fotostock

Engineering cars is hard. Engineering semis is harder.

Now that you've seen some of the players in the BEV semitruck space, let's return to the two items I mention earlier in this section.

In the case of the Tesla Semi, the truck was announced during an event in 2017, with a promised delivery date of 2019. (If I had to guess, which I do have to guess, I would say that the anticipated 2019 date was announced with the expectation of an improved battery being ready by then. In terms of the motors and other vehicle components — the prototype Tesla Semi has four along its rear axles — that engineering nut has already been cracked.)

However, as with several other Tesla projects <looks askance at Full Self Driving (FSD), Cybertruck, and Roadster>, not exactly achieving announcement dates is the rule, not the exception. The Semi now isn't expected to enter volume production until 2023 — at the earliest. That's four years after the announced launch, and that's still a best case scenario.

Trust me: It's all about the battery.

Though Tesla has been pushing back the anticipated production date of its Semi, it's a forgivable miscalculation.

The Tre from Nikola Motors, meanwhile, was infamously rolled down a hill(!) at the direction of then-CEO Trevor Milton(!!) to demonstrate the truck's performance "in motion" (you, too, should be using air quotes when reading this sentence) rather than being driven under the power of its own drivetrain.

I do want to point out that Nikola has since remedied this, and Milton is now staring at the unemployment line, along with possible time in the hoosegow for having committed fraud, but the larger-picture issue remains. It is this:

> Electric semis *will* play a massive role in the global shift from oil and other fossil fuels to electric power. Deploying fleets of electric semis has the potential to make a tremendous impact on both greenhouse gas emission goals and the bottom line of the companies that order them. For this last reason alone, electric semis are sure to be commercially deployed the second that battery technology becomes commercially viable.

Though many challenges still lie ahead, significant progress has already been made. Full conversion is just a question of when.

But heavy-duty trucks might not even be the venue where electric trucks are introduced to the general population.

The very first electric truck you see? It's likely to be an electric van.

Delivering the Goods (with Electric Delivery Vans)

Another significant and easily overlooked segment of the world's transportation fleets is delivery vans and trucks. Think of all the mail delivery trucks sweeping up and down neighborhood streets every day, not to mention all the UPS, DHL, and FedEx delivery vehicles, and of course the ubiquitous Amazon smile painted along the sides of hundreds upon thousands of delivery vans.

Compared to the typical passenger car, these kinds of last-mile vehicles are driven far, far more frequently, and rack up many tens — no, hundreds — of thousands of miles during their lifetimes.

In the logistics/delivery business, *last mile* refers to the point at which the package finally arrives at the buyer's door. Last mile delivery is both the most expensive and time consuming part of shipping something to your door.

However, because they're big and not terribly fuel-efficient (they're probably as fuel-efficient as they can be), and because they're being driven around all day every day, they're responsible for a lot of carbon emissions. By EPA estimates, these medium- and heavy-duty trucks account for about 25 percent of greenhouse gas emissions from the US transportation sector. (You can check my figures at `www.epa.gov/greenvehicles/fast-facts-transportation-greenhouse-gas-emissions`.)

But here's the thing: Decision-makers at places like Amazon and UPS aren't intentionally underwriting an effort to slowly broil our planet — it's hard to deliver fresh groceries when crops fail, after all. It's just that their interest in the bottom line likely supersedes their interest in greenhouse gas emissions.

Fortunately, electric delivery vehicles offer a strong value proposition over ICE vans. As with electric passenger cars, the initial costs can be significant, but for many of the firms interested in a fleet of delivery vehicles, the capital expenditure investment pays off in the form of dramatically reduced operation expenditures. It simply costs less to operate an electric van per mile than it does an ICE van.

Capital expenditures involve the initial cost of equipment needed to run or expand a business. Building a new factory and buying the machines for that factory are all capital expenditures.

Operating expenditures are the ongoing costs of running the business. In the matter of a delivery van, the purchase price of the vehicle is a capital expenditure, and the cost of fuel and maintenance is an operating expenditure.

So, although an electric delivery van may have neither the sex appeal nor the 0-to-60 times of luxury electric sedans, big entities have plenty of financial incentive to put things into trucks and drive those things to houses and businesses to make the switch. And so it follows that several companies are trying to make a product to fill that market need.

One newsworthy order has come from Mr.-size-Medium-shirt-on-a-size-Large-chest, Amazon's Jeff Bezos. In 2019, Amazon, which owns a significant stake in automaker Rivian, placed a 100,000-vehicle order for Rivian to go make it some electric vans.

In 2022, though, Amazon placed another order for electric vans, this time from Stellantis, which means that the EV vans would technically carry the Dodge Ram badge.

Whether from Silicon Valley's Rivian or from Amsterdam's Stellantis (Amazon has previously ordered thousands of gas-powered trucks from Stellantis), the online retailer should start delivering its wares to your door using EVs starting in 2023.

With that, the next few sections describe some of the players in the electric medium-duty vehicle space.

BrightDrop

General Motors is rapidly expanding its electric options, including two new delivery vehicles from its BrightDrop subsidiary: the EV600 and the EV410. The two tout 600 and 410 cubic feet of cargo space, respectively, and both offer a claimed range of up to 250 miles. (See Figure 12-10.)

Canoo

Canoo has already unveiled plans for three separate models: a minivan-like lifestyle vehicle, a pickup truck, and its MPDV (*m*ulti*p*urpose *d*elivery *v*ehicle) line. (See Figure 12-11.) The latter product line now entails two models: the MPDV1 and the MPDV2.

Rivian EDV

With Rivian's assistance, Amazon has announced its duo of electric delivery vans, newly dubbed the EDV line. The first to be delivered is the midsize EDV 700, with a smaller EDV 500 coming later. The numeric designation on the models denotes the size of the cargo hold — 700 and 500 cubic feet, respectively. (See Figure 12-12.)

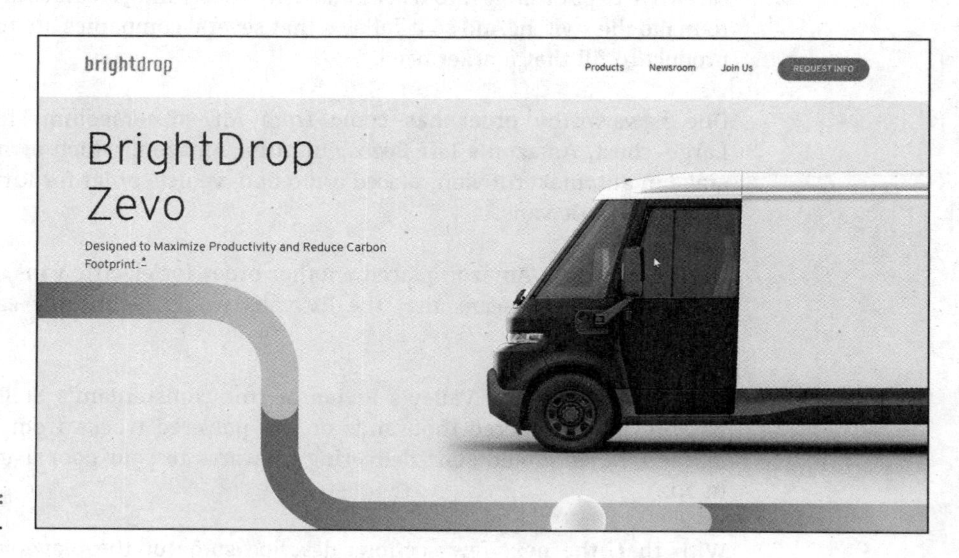

FIGURE 12-10: The BrightDrop.

FIGURE 12-11: The Canoo MPDV delivery van.

FIGURE 12-12: The Rivian.

mark reinstein / Shutterstock.com

Workhorse

UPS delivery vans are easily recognizable by their brown color scheme, and in 2020, the Workhorse Group began building an all-electric delivery van, called the C1000, for this flagship customer. (See Figure 12-13.)

FIGURE 12-13: The Workhorse C1000.

Additionally, Workhorse is reportedly beginning work on delivery vehicles for the US Postal Service, which could result in over 150,000 USPS mail being replaced by electric varieties. The Workhorse Ranch is located in Union City, Indiana.

Keeping things in perspective

This book is targeted to the individual vehicle consumer. These consumers shop for and drive either passenger cars or light-duty trucks (that is, pickups).

The topic of medium- and heavy-duty trucks is more of a concern with fleet purchases. The potential savings from people hauling stuff or using electric is certainly relevant, but the average reader of this book will have no say in the city's next purchase of 10 electric buses, FedEx's next purchase of 100 electric delivery trucks, Coca-Cola's next purchase of 1,000 semis, or the US Postal Service's next purchase of 10,000 mail carrier trucks.

So, although the topic is quite interesting and timely, from both the financial and environmental perspectives — maybe you'll invest in a company that manufactures last-mile delivery vehicles or decide to vote with your dollars by shipping more often with the company utilizing EVs — but, otherwise, it's something most of us will never directly interact with.

GOING POSTAL

The USPS has recently (Feb of 2022) ordered 150,000 new mail trucks from a company called Oshkosh Defense, and these trucks are gas-powered rather than electric, alas. The EPA recommended that the USPS pause the contract and look more closely at EV postal carriers. The proposed new ICE trucks are apparently just .4 miles per gallon more efficient than the trucks being replaced.

Not 4 miles more efficient. Point 4 miles.

However, Postmaster General (PG) Louis DeJoy — who personally contributed over $1 million to Trump's reelection campaign and was then given the PG position because he was clearly the right person for the job, and clearly *not* because there was any sort of *quid pro quo*, no sir, and how dare you even think that — went forward with the ICE order anyway.

Electric truck maker Workhorse (see above) has sued the USPS over the matter.

As you can guess, there are several petitions floating around to try and get the USPS to reconsider, including one from the United Auto Workers (UAW) union. The UAW argues that these delivery trucks should be both electric and made with union labor, neither of which are happening with Oshkosh.

(The whole episode has another weird twist in that Oshkosh Defense workers in Wisconsin — UAW Local 578 — are protesting their own company, who it must be restated *won the darn contract* from DeJoy. Except that Oshkosh is apparently planning to make the postal vehicles out of state (Spartanburg, South Carolina to be exact) using non-union workers, so the UAW at Oshkosh Defense HQ in Wisconsin is now pressuring their own company not to go forward with their plans.)

In any event, someone please send DeJoy a copy of this book. Come on, dude. You've got kids and (I assume) grandkids . . . do the right thing.

I Sing the Lawn Mower Electric

I know the title of this book is *Electric Vehicles For Dummies*, and the electric vehicles that everyone pictures when they hear it are the kind you sit in to travel from point A to point B.

But lawn mowers, too, have four rubber wheels and a motor, and some of them you can ride. That said, it's best not to wind me up about the topic of lawn mowers.

In fact, don't get me started about lawns — they're dumb, wasteful, and almost shockingly bad for the planet. Lawns are the single largest irrigated crop in the United States, and — for what? All that time, money, and water, and for what, exactly? Apparently, we're a nation of millions of gentlemen farmers raising a crop that we have to . . . bag up and throw in the landfill? Why on earth do we do this?

When my family moved from Kansas City to San Francisco, we moved to a place that had no lawn. And it's been liberating, like a veil has been lifted from my eyes, allowing me to see things in ways I haven't before.

Human beings don't *require* a lawn! They can be perfectly happy without dedicating two or three hours of every Saturday to cutting grass or edging the driveway or applying fertilizer or picking up twigs or spraying for dandelions or crabgrass or raking leaves for two months in the fall or seeding over brown spots. Ugh. Ugh to it all. Rather than tend to their lawns, humans can use that time to ride their bikes or take a walk with their dogs or read a book or watch a movie or bake bread — or just do nothing at all! It is possible! When I look back on my years of lawn care, I actually feel shame that I've wasted so many hours on something so utterly trivial.

OK. I feel better now. I'll keep the focus from here on the equipment used for lawn care. (Which of course you don't need if you have no lawn.) So, if you must have a lawn (you don't), but if you must (seriously, you don't), please make the first thing with four wheels and an electric motor that you buy an electric lawn mower.

With a little digging, you can find the studies that back me up, but just trust me when I say that none of them is complimentary when it comes to the greenhouse gases tied to mowing a lawn with a gas mower. By one estimate, running a lawn mower for an hour emits as much greenhouse gases as a 100-mile trip in an ICE car. Meanwhile, the EPA estimates that a single gas-powered lawn mower is responsible for 11 times as much pollution as a gas-powered car, or about as much air pollution as 43 new automobiles driven 12,000 miles per year. (You can check those numbers at `www.epa.gov/sites/default/files/2015-09/documents/banks.pdf`.)

Please. Kill your lawn before it kills the planet.

Or, at the very least, ditch your ICE lawn mower for an electric model. You have lots to choose from now, thanks to new battery tech that's making them more powerful, longer-lasting, and — like their brethren, the electric-powered car — cheaper to maintain over the long haul.

No matter which model you buy, it will look something like the one you see in Figure 12-14.

FIGURE 12-14: If you must mow, electric you shall go.

tab62 / Adobe Stock

Here are a few models worth your consideration, should you want to continue spending Saturday afternoons tending to a small plot of grass that could just as easily be given over to zero-maintenance plants that are native to the area:

- EGO Power Plus 56V Brushless
- Ryobi One Plus HP 18V Brushless
- Hart 40V Cordless Brushless
- Greenworks G-MAX 40V
- Kobalt 80V Max Brushless

Each of these electric mowers is a better choice than a gasoline mower in terms of noise, cost of operation, and greenhouse emissions.

And don't get me wound up about leaf blowers. I said, don't —

> ***A public service announcement from the editors at Wiley:*** We at Wiley, in our great editorial wisdom, have decided to excise certain comments about leaf blowers, out of respect for your time and sanity. Let's just say that Brian is even less a fan of gas-powered leaf blowers than he is of gas-powered lawn mowers and leave things there. We at Wiley do encourage you to research your own electric lawn mowing and leaf blowing options the next time you're considering replacing old lawn equipment. And should the subject of leaf blowers ever come up while in conversation with Brian, we encourage you not to make direct eye contact.

» Totting up benefits of biking

» Categorizing the various ebike models

» Looking at who's in the ebike game

Chapter 13

The Best Car Is No Car: Electric Bikes and the Morning Commute

I have a request: Please consider *not* getting an electric car.

Should I have led with that? Maybe so. I'm spending a massive amount of time and energy extolling the upsides of electric vehicles, and here I am in a late chapter advising you not to bother. Talk about burying the lede.

Look: I don't pretend to know what's best for your transportation needs, especially given how most of the US has been built after World War II. Over the course of just a few short generations, most of the US (and Canada, but the development pattern most certainly started in the postwar United States) has been conducting an experiment that has no precedent in human history. We've fundamentally altered how we build and inhabit our cities, and many of us don't realize that it wasn't always this way — that it was indeed altered, and that this 75-year experiment runs counter to how humans have inhabited built environments for thousands of years.

Many of us, especially many of us who are interested in electric personal transportation, now live in commuter suburbs that are lined with grids of collector and arterial roads (roads that are neither street nor road and that are brutal places to exist outside a car) that are in turn connected to downtown employment centers by way of massive freeways that are anything but free. When I look at aerial views of suburbia, I'm reminded of the bars of a prison; many of us are born into a landscape of car dependency that begins with our first trip home from the hospital.

I lived in such a place growing up: Kansas City, a place that has *more pavement per capita* than any other city in North America. I just *wrote* that sentence, and still I had to stop and reread it.

Who knew? Certainly not I, when — as a 14-year-old — I started saving money for a car so that I could get myself to school (and work).

And I live in such a place now: Fremont, California, where neighborhood streets that kids use for biking and walking to school are engineered like drag strips. Because of this, it's a place where drivers treat them as such, something I also mentioned in a sidebar in Chapter 3. (See Figure 13-1.)

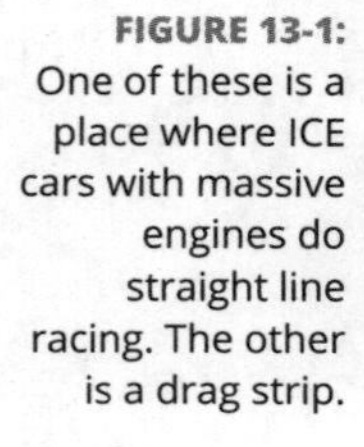

FIGURE 13-1: One of these is a place where ICE cars with massive engines do straight line racing. The other is a drag strip.

Brendan / Adobe Stock

I mean, why take into consideration the safety of a child when you have a yellow light to beat?

We humans are now starting to see the results of this multigenerational experiment, and these results point to the experiment being a colossal failure: a failure economically, culturally, environmentally — and as a matter of public health. If you're thinking of not even reading this section because you'd never in a million years consider riding a bike down the same strip of pavement you use to drive to work, you're hardly alone. I've been cycling for more than 15 years and I don't feel 100 percent comfortable even riding in a bike lane where cars and delivery trucks are zipping past at 40 and 50 miles per hour.

Oh, and all of this is before we acknowledge the racism that's deeply embedded in every highway cutting through a city, and in every suburban cul-de-sac (a cul-de-sac is a publicly funded and privately used driveway for the people who can afford to live there). For those who want to tell me that a highway can't be racist, let me just pass on a reading recommendation: Eric Avila's *The Folklore of the Freeway: Race and Revolt in the Modernist City*, published by the University of Minnesota Press.

Or consider this image from the Kansas City Public Library (see Figure 13-2), showing space cleared for the "downtown loop" (the interstate highway, in other words) during the 1950s and ask yourself whether or not any people were displaced in order to bulldoze that strip of city?

FIGURE 13-2: Spoiler alert — there were people displaced.

The Kansas City Public Library

There is virtually no measure of the US pattern of the suburban development experiment that can be called a success, except perhaps in that it has excelled at making numbers tick up in the bank accounts of the land developers and speculators who've helped build out all of this sprawling mess. I guess it's also good if you're in the lawn mowing business (which, as a teen, I was).

How to undo all this? That's the subject of another book.

How can *you* take some small steps to undo all this? By considering an ebike.

Making the Trip by Bike

Despite the fact that for generations we humans have been born into car dependency and have little say over its continuation (see the diatribe that opens this chapter), about *half* of all car trips are three miles or fewer, and the majority of car trips are fewer than six miles. Let those numbers sink in.

Don't get me wrong: I *love* electric cars, and I think they are better for humankind and that they hold the promise of a more equitable, egalitarian, less car-dependent future (or at least a less car-ownership-dependent future), and there's no way I could have poured in all the hours necessary to create this book — the many late nights and weekends hovered over the keyboard — unless it was truly a labor of love.

Also, yes — I own an electric car and have no plans to sell it anytime soon. But I also own an electric *bike*, and for my daily work commute, I use that bike almost as much as I use my electric car.

My work commute: 6.1 miles from door to door, making this daily commute one of the majority of car trips made every day in the US. (My morning commute to high school, by the way — a trip I made every weekday by car, starting at age 16? It was 3.5 miles.)

The current split on work commuting is about 50-50. And, on the weekends, all grocery and coffee runs are made using the bike, not the car, unless the weekend trip involves carrying something heavy or bulky or canine.

So, clearly, I'm a great guy for using a bike for the (slight) majority of my daily travels, and also for rescuing a shelter dog (see Figure 13-3), and also being able to convince the Wiley editors to include a section about ebikes, which ironically might be the most valuable few pages for a plurality of readers in a book full of advice about saving money by buying an electric car.

The question, however, is whether a bike — or, specifically, an ebike — is a good fit for you or your family, and what kinds of factors might go into your ebike decision. As you'll see in a moment, even though I've been an avid cyclist for about 15 years, having an ebike has opened up new possibilities, and it has been especially effective at replacing short car trips.

The next few sections look at five ways an ebike can prove beneficial.

FIGURE 13-3: Alexander Remington — he's young, scrappy, and hungry. Would probably bike if he could, but does happen to like car trips to the park.

Improving fitness

In terms of overall fitness, do you know what pedaling an ebike is like? It's like pedaling a bike — just a little less perceived effort (and a little more speed) is all.

In other words, the regular use of an ebike will categorically improve your fitness level. Numerous studies back up this statement, including ones that show that the hearts of ebike riders are working at about 90 percent of the level of non-assisted bikes. (Check out the stats at `https://formative.jmir.org/2019/3/e13643/`.)

But do you really need a study?

It's common sense stuff: In a car, you're sitting in a box and probably increasing your cortisol levels as you wait at stoplights and develop road rage about the traffic on the highway. (You're not stuck in traffic — you *are* traffic.) On a bike, you're outside, pedaling a machine, expanding your lungs, and improving the efficiency of your cardiovascular system.

Improving mentals

Speaking of fitness, let's not forget to mention the mental fitness that any form of cycling imparts.

I can tell you anecdotally from 15 years of doing this that cycling is a good way to be alone with your thoughts for many minutes, or even hours, of the day. It's a chance to think, to process a day's or a week's events. I've laughed and cried while riding a bike; I've given my subconscious time to chew on a problem (like how to structure a book on electric cars) and have had some amazing epiphanies while biking that have helped me unstick a problem, either personally or professionally.

I'm not sure why this happens. I've heard psychologists liken it to being in the shower and allowing your brain to relax. I've heard some compare cycling (or walking or running) to REM sleep — that there's something about the left-right-left-right motor signals that happen during cycling that mimics the left-right-left-right eye movements that occur during diurnal (REM) sleep periods.

The bottom line is that cycling has proven — by way of scientific study and by way of my own experience — to impart benefits like reduced anxiety and stress and like greater overall happiness.

Science also says that regular cycling will help you sleep better, too.

Enhancing your ability to explore new places (and more often)

Would I ride my bike to work if it weren't an ebike?

Yes, but probably not as much. When I ride my road bike, the 12 miles of total commute are just far enough that I can feel the fatigue the next day. Plus, the non-assisted ride takes a few minutes longer and I arrive at work a little sweatier.

Ebikes enable you to ride further and with less effort. I can ride 6 miles to work in my business casual clothes and not worry about finding the work showers or packing a change of clothes. I can ride home that evening and not feel the day of riding in my legs.

When you're not riding to work, the ebike can take you up more challenging climbs, or just simply let you keep those pedals spinning longer.

And more often. Other research about the impact of electric bikes has shown that ebike owners head out more frequently than non-assisted cyclists. The reasons seem clear: An ebike gives you more confidence about where you can go, and it makes you more confident about how you'll feel after you arrive.

Having access to faster and cheaper transportation

As should be obvious if you're reading the entirety of this book, riding an ebike is less expensive than driving a car. What's more, ebikes are usually *faster* for most short- to medium-length journeys.

Ebikes don't require licensing. They don't require insurance; they don't even require a battery — you can happily pedal an ebike with a dead battery or while leaving the battery on the charger.

Add this benefit to the cost of a car and swapping out a car for an ebike can put tens of thousands of dollars per year into your pocket. Even factoring in the cost of the bike and the electricity to charge the battery, an ebike's operation cost per mile is a fraction of what it costs to drive a car for urban trips.

If you're driving a mile or two, the door-to-door time of a bike and car is probably equivalent, especially considering that you don't have to park and walk through a parking lot. When I go grocery shopping, I lock the bike within a few feet of the front door.

And once again returning to my commute, it takes me between 15 and 20 minutes to get there by car. It takes about 20 or 22 minutes by bike.

Putting money in your pocket through tax incentives

When it comes to the cost of an ebike, you might think that there'd be a significant price premium because of its motor and battery. But you might be surprised at how cost-competitive an ebike can be.

Plus, many municipalities have incentives that put ebike ownership even closer on par with a similar non-ebike. Before purchasing, it's worth a quick search to find out what tax incentives are available in your country, state, county, or even city.

Here's an example from Santa Clara County, California. Santa Clara has a population of almost 2 million and contains the cities of San Jose, Cupertino, Mountain View, and Palo Alto.

> "Customers of Silicon Valley Power are eligible for a 10% rebate on the price of an ebike, up to $300. Customers in their Financial Rate Assistance Program can receive an extra $200 per bike."

Meanwhile, in Austin, Texas:

> "Austin Energy is offering rebates for both its residential and commercial customers. Up to $300 rebates for individuals and up to $400 for fleets of new bikes."

And, as you might guess, ebike manufacturers want to make sure you're well informed about available incentives, because an extra couple hundred dollars can help tip a buying decision in their favor.

I found out about each of these two incentives at the website of the ebike manufacturer Aventon, and, more specifically, at this URL: `www.aventon.com/blogs/aventon_bikes/ebike-rebates-and-incentives-across-the-usa#`.

But again, an online search should help you track down available incentives in your locale.

Meanwhile, across the proverbial pond, United Kingdom citizens can leverage the Cycle to Work scheme, which includes purchases over £1,000. In Scotland, you can also take advantage of an interest-free loan to buy an ebike of up to £6,000 using a program sponsored by Transport Scotland.

Now that you've been exposed to the benefits of frequent ebike use, it's time to delve into the shopping aspect. You might think that a bike is a bike: two wheels, a chain, and handlebars. Or that an ebike is a bike just with a small electric motor.

But there's more to it than that, and it can quickly become overwhelming. One of the first things that can help you cut through some of the confusion is understanding what is meant when someone refers to an ebike *class*.

Looking at the Three Classes of Ebikes

If you think there are a lot of possibilities when shopping for electric cars, just wait until you start searching for ebikes.

First of all, because they are much less complex (and much less expensive) to build than a car, you have many more ebike manufacturers to choose from. You just don't need as much capital to build a website and start selling $2,000 ebikes, compared to what's needed for $60,000 e-cars.

REMEMBER

The classes I'm talking about here are *legal* classifications. They ostensibly govern where they can be sold and where they can be ridden, although it varies from jurisdiction to jurisdiction.

Three classes of ebikes are now for sale in the market:

- **Class 1:** Class 1 ebikes have a top speed of 20 miles per hour. Additionally, the electric motor works only when the rider is pedaling.

  A Class 1 ebike can be ridden on bike paths and bike lanes that are used for traditional bike traffic.

- **Class 2:** Class 2 bikes also have a top speed of 20 miles per hour. What separates Class 2 bikes from Class 1 bikes, however, is that Class 2 bikes have a throttle — you push a button, usually located on the handlebars, and the bike takes off, up to 20 mph. The electric motor still assists when you're pedaling; it's just that, with a Class 2 bike, you don't *have* to pedal.

  REMEMBER

  Some Class 1 ebikes, like the VanMoof S3/X3, have a throttle that offers a boost in power, but a Class 1 throttle kicks in only when you're pedaling. You don't have to be pedaling hard — just barely turning the pedals is all that's required for the throttle to do its thing.

- **Class 3:** Class 3 ebikes can pedal-assist to a top speed of 28 mph — as a Class 3 ebike owner and commuter, I know that's *plenty* fast — in fact, probably *too* fast for comfort in most residential neighborhoods. (I personally feel like I don't have enough reaction time should a car back out of a driveway or a dog decide to chase a squirrel across the street.)

  In most places, throttles are allowed on Class 3 ebikes as long as they only work up to 20 miles per hour.

  Class 1 and 2 ebikes are legal on any paved surface that a regular bike is allowed to operate. (In most municipalities, you're not *technically* supposed to ride a bike on the sidewalk unless you're under 16.)

  In most states, you can pedal a Class 3 ebike on either regular road lanes or bike-only lanes. You're not supposed to take a Class 3 bike on bike paths or on multiuse trails shared with pedestrians, like in a park.

  That said, I've seen plenty of Class 3 bikes on shared paths, their owners opting for one of the lower levels of assistance, topping out at 10 mph, or maybe 15 mph, of assistance. Though I, of course, would not advise anyone to break local ordinance, I have yet to see a policeman with a radar gun pointed at cyclists on the Alameda Creek Trail. I mean, if I took my Trek road bike around a park trail shared by pedestrians, I would never pedal at full speed.

In any event, now that it's clear that most of today's new ebikes are either Class 2 or Class 3, let's look at where you might start some of your comparison-shopping.

Scouting Out the More Notable Ebike Companies

You have many, many ebike manufacturers and retailers to choose from today, far more so than manufacturers of electric cars. With that said, let me pass along in this section a few that made my short list when deciding on my first model.

VanMoof

www.vanmoof.com/en-US

Originating in Amsterdam, VanMoof makes sleek, stylish ebikes targeted to urban commuters. (See Figure 13-4.)

FIGURE 13-4: A VanMoof.

"Truly city-proof, VanMoof bikes shrink long commutes, scare off bike thieves, and amplify your pedal power. No wonder cities around the world are scrambling to build more bike lanes."

Most of VanMoof's bikes are Class 2 bikes, which is probably the main reason I didn't choose a VanMoof for my first ebike. But it now makes a 31 mph model called the VanMoof V, which is now being added to my letter to Santa.

Gazelle

www.gazellebikes.com/en-us/about-gazelle

Another manufacturer that was founded in Holland — it's almost like the Dutch are known for cycling — Royal Dutch Gazelle "... has been making bikes for more than 130 years in its factory in Dieren," with a focus on safety, comfort, and design.

Gazelles are great-looking machines, with the Ultimate model a particularly good example of the Gazelle aesthetic. (See Figure 13-5.) It's a Class 2 ebike that features top-notch components, an integrated battery in the frame, hidden cabling, and powerful Bosch eBike Systems.

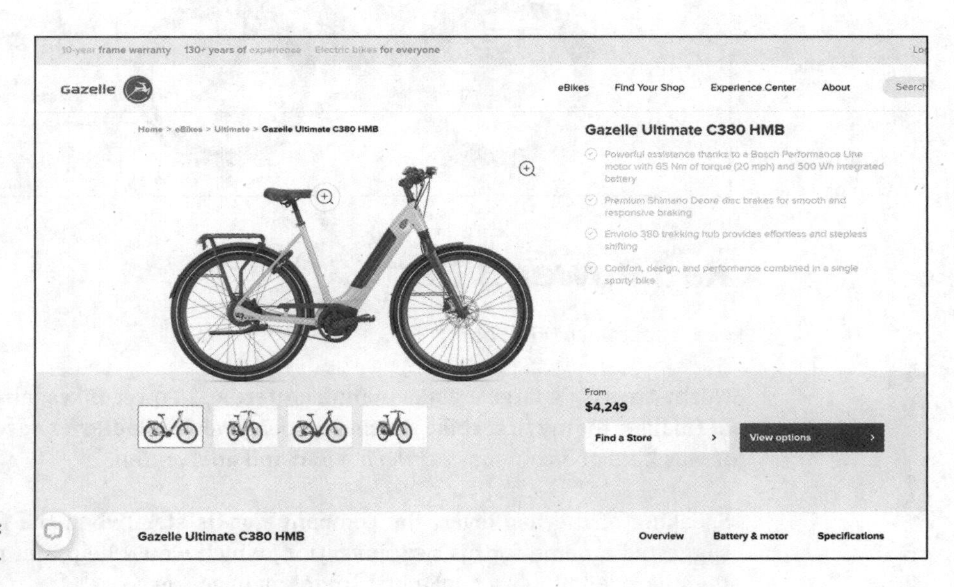

FIGURE 13-5: A Gazelle.

Aventon

www.aventon.com

Aventon is a fast-growing ebike manufacturer and direct seller that makes budget-friendly commuter bikes and also offers several choices of Class 3 ebikes.

It was this combination of not wanting to break the bank when dipping my foot into the ebike waters, along with being able to ride at 28 mph, if necessary, that ultimately made me decide that Aventon was the right choice for my first ebike.

Aventon's Pace, Level, and Soltera are all excellent commuter choices. (Figure 13-6 gives you a peek at the Level.)

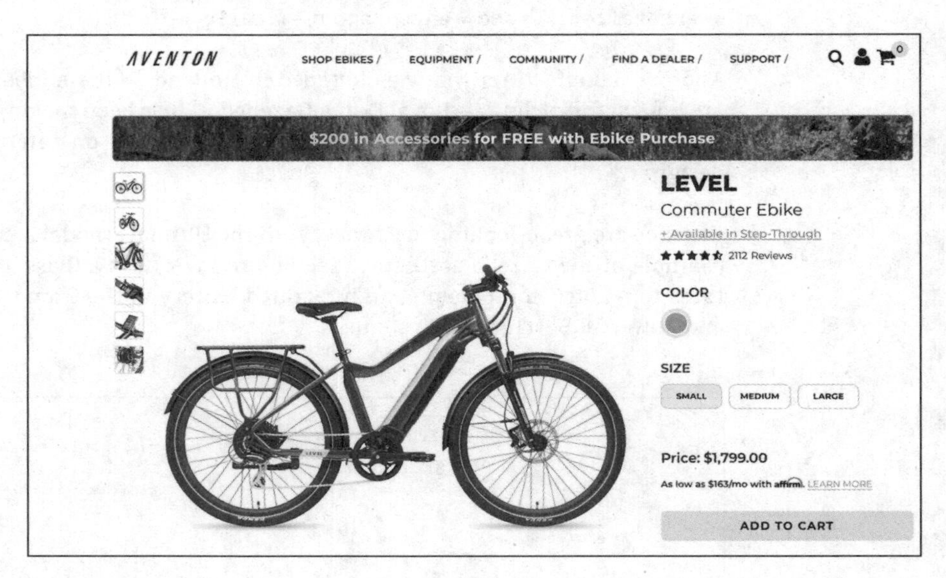

FIGURE 13-6: An Aventon Level.

Rad Power Bikes

`www.radpowerbikes.com`

North America's largest ebike manufacturer, Rad Power Bikes, also made my list of finalists for my first ebike purchase (specifically, a RadRover 6; see Figure 13-7). It was kind of a coin toss between a Rad and an Aventon.

Speaking of the RadRover, the company got its start when the founder's mom suggested a name for his new invention, which would help him make the hilly, 15-mile ride to high school a little more manageable.

My wife and I have recently gifted a RadWagon, which is an electric cargo bike with a 350-pound. weight capacity, to a new dad so that he can make grocery runs with his daughter.

FIGURE 13-7: The RadRover 6, from Rad Power Bikes.

Bird Bikes

`https://bike.bird.co`

Yes, *that* Bird — the one that is probably more famous for its fleet of electric scooters, which is yet another great way to get around in an urban environment without the physical or carbon footprint of a car.

When you visit the Bird ebike online shop, you might think you've been redirected to the VanMoof website, but they are indeed different entities. Bird Bikes come in either an A-frame (step over) or V-frame (step through) variety, are reasonably priced, and have cool features like standard splash guard fenders and an integrated backlit handlebar dash display. (See Figure 13-8.)

We recently bought a V-frame for our daughter for commuting around her college campus, and she loves it, which, if you knew how cycling-averse our daughter was before the ebike, makes *She loves it* three words that are a remarkable thing to write in sequence about a bicycle.

An ebike makes all the difference.

Trek

`www.trekbikes.com/us/en_US`

I'm mentioning Trek here because I've test-ridden a Trek ebike (the Verve+) and quite enjoyed it, and also because I've had a lot of Trek brand loyalty over my 15 years of cycling. I just trust what they make; I love their bikes — full stop.

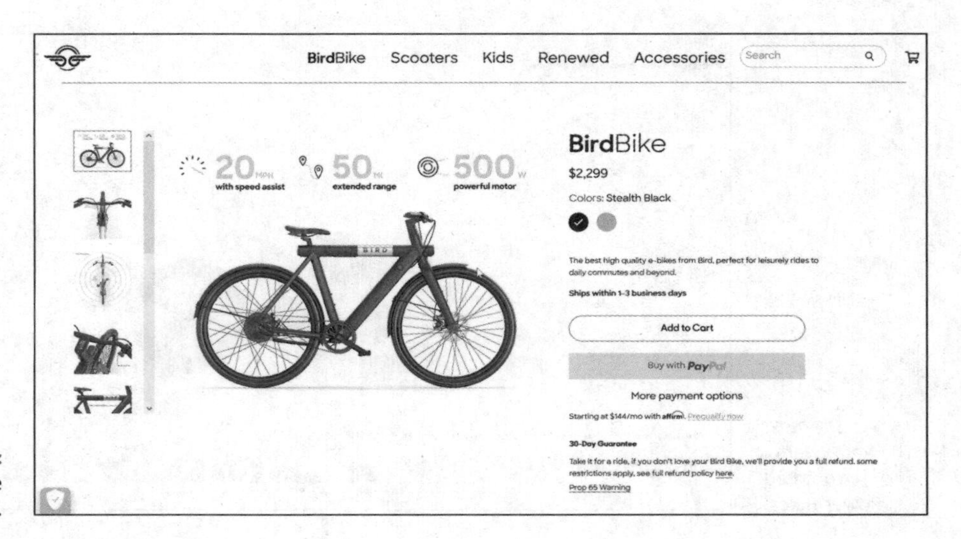

FIGURE 13-8: The Bird Bike A-frame.

What made me hesitate with a Trek commuter bike was a) the price and b) the fact that their current models are Class 2. That said, if I were shopping for an electric mountain bike, I'd likely go with something like a Trek Powerfly. (See Figure 13-9.)

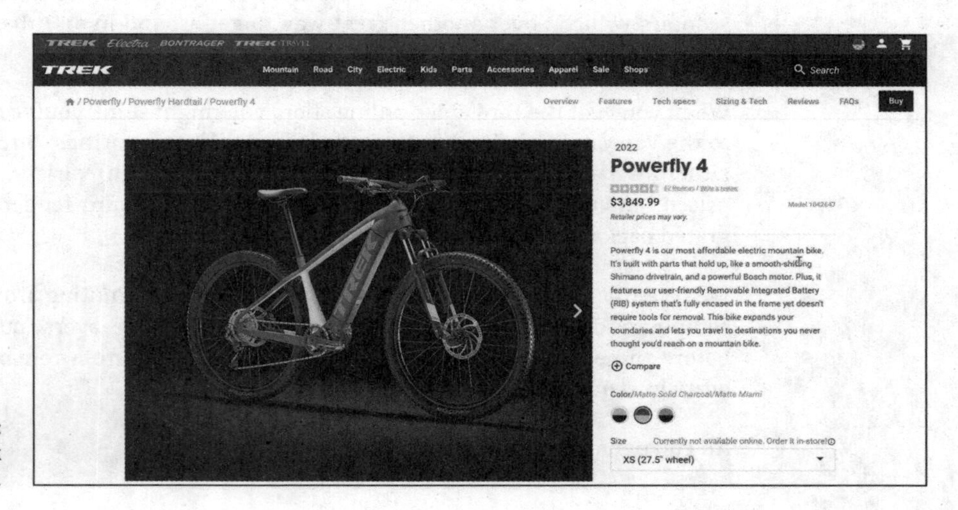

FIGURE 13-9: The Trek Powerfly.

What I'm saying is that you won't go wrong with a Trek, and it's usually a bit easier to get Trek bikes serviced through their many retail locations. A direct-to-consumer brand like Aventon or Rad can require a bit of DIY savvy from owners. If you want to only turn the pedals and let others turn a wrench, I strongly urge you to consider the bike *shop* you're getting the bike from.

Specialized

www.specialized.com/us/en

The last mention here goes to Specialized, which, like Trek, is a reliable bikemaker that offers both a wide variety of ebike options and a generally strong network of bike shops where you can go for the occasional tune-up, repair, and other services.

Specialized has been making bikes since 1974, and here's what its marketers have to say about the company's more recent foray into electric cycling:

> "When we developed our first electric bike in 2009, we knew it had to ride and feel like a great bike first — something we'd really love to ride ourselves. We wanted you to feel like you were on a "regular" bike, but that somehow, you'd grown superhero legs . . . It's You, Only Faster."

Both the Turbo Vado and the Turbo Como are Class 3 ebikes that are high on my list of possible commuter bikes to purchase next. (For a peek at the Vado, see Figure 13-10.)

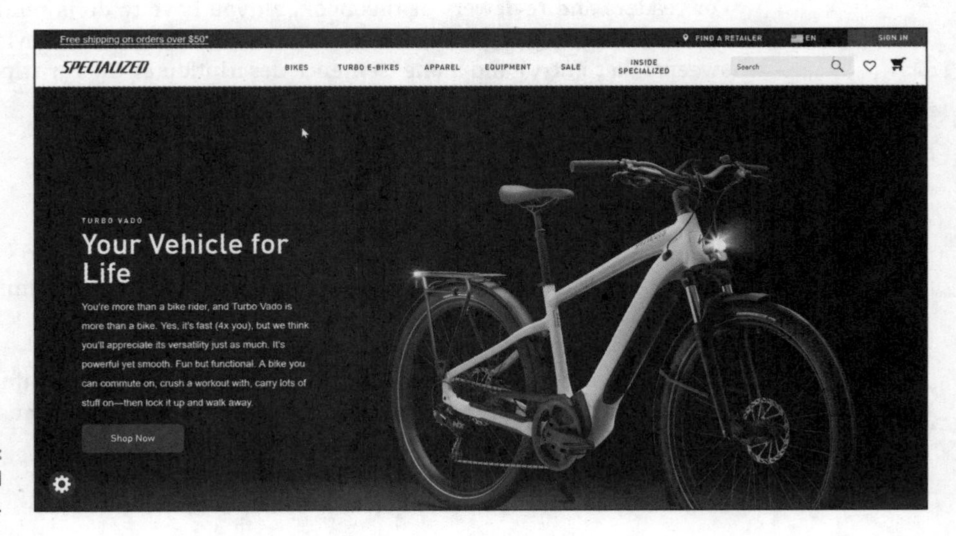

FIGURE 13-10: The Specialized Vado.

I say "high on my list" of ebikes to purchase next because I'm raffling off my Aventon to a reader, and I will continue to do so. More about that next.

If you're curious, *Bicycling* magazine (www.bicycling.com) tested current ebike offerings and rated the Aventon Level the overall best bike. It's a Level 3 bike that comes with front shocks and disc brakes and can be purchased for a reasonable US $1,799. (For all the standings, check out www.bicycling.com/bikes-gear/a22132137/best-electric-bikes.)

The final word here is that you can't go wrong with a bike ride that replaces a short car trip. You'll be healthier.

It's especially fantastic if someone buys the ebike for you.

Hey! I'm Giving Away Ebikes!

Really.

I believe in the power of bikes so strongly that I'm offering an incentive that I'm backing with my own resources.

For readers and reviewers of this book, all you have to do is send a link to your Amazon review (yes, you can mention that you submitted a review as part of a sweepstakes entry) and a one-sentence description of the car trip that your bike will replace.

Use the form on the book's companion website to enter:

www.brianculp.com

Every quarter, I'll gift one winner an Aventon Pace 500 (or similar) commuter bike.

And, hopefully, this one-at-a-time giveaway will lead to something much bigger. I'm planning to use the website to recruit others to support this mission of replacing cars with bikes, one trip at a time.

So, as they say in the world of electric vehicles: Watch this space.

It should prove endlessly fascinating to do so.

HOW TO HELP

My wife and I donate yearly to two cycling charities, and we encourage you to consider these in your giving plans: One is People for Bikes (www.peopleforbikes.org), which performs a host of advocacy and educational work at the state and local levels.

Among tons of other amazing resources, People for Bikes has a state-by-state compendium of electric-bike laws because electric-bike rules can vary considerably, depending on jurisdiction. Did you know that Massachusetts requires an operator's license for ebikes and that you're supposed to be 16 to ride one? Well, you would if you familiarized yourself with the People for Bikes website. (Fortunately, most states are not as problematic for ebike riders as Massachusetts.)

Check it out here: www.peopleforbikes.org/electric-bikes/state-laws.

The other charity worth reading about is World Bicycle Relief, which you can find out about at Trek's website: www.trekbikes.com/us/en_US/wbr-donate.

World Bicycle Relief uses donations to buy and deliver Buffalo bikes, which help people more easily attend school, receive healthcare, and deliver goods to market.

5
The Part of Tens

IN THIS PART . . .

Track the countries in the forefront of EV technology

See which companies are making an EV impact

Take a look at Brian’s crystal ball

IN THIS CHAPTER

» Looking to Europe

» Checking out the Americas

» To infinity — and beyond

Chapter 14

Ten Countries Leading the Way — and What They're Doing

Here's a fun fact: According to the International Energy Agency, global electric car registrations grew by 41 percent in 2020 while car sales overall *dropped* by 16 percent.

Granted, the world experienced a pandemic during that same time frame, but the trend is clear: ICE sales down, EV sales on the rise. The bottom line is that in 2020, this 41 percent rise in EV sales meant that someone drove electric car number 10 million home from a dealership.

The countries described in this chapter are where most of those 10 million EVs are being deployed.

Norway

Norway is absolutely crushing it, electrically speaking, leading the globe in the switch to electric transportation. As reported by the Australian website The Driven (`https://thedriven.io`), in January of 2022, 83 percent of all new passenger vehicles registered in Norway were electric. And not hybrid electric — *fully* electric. (One was a hydrogen-powered car, apparently.)

Add the 83 percent of all registrations in January 2022 to previous month's totals, and roughly 25 percent of every car in Norway today is electric. (See Figure 14-1.)

FIGURE 14-1: Norway: Showing the way, leading the way.

Photo credit: Grigory Bruev / Adobe Stock

In large part, this is because the people who govern Norway have made electric transportation a top priority and have not shied away from the task of leadership when it comes to the policies that make the transition easier for its citizens. For example, EVs are exempt from all nonrecurring vehicle fees. They are exempt from 25 percent of their VAT (Value Added Tax) on purchase and have lower public parking fees and toll payments. EVs are even given access to bus lanes.

And those are just the incentives that are still active. Several others have already been phased out, like the 50 percent reduction in company car tax or the free municipal parking that was in effect from 1999 to 2017.

Add it all up and driving an EV in Norway is a no brainer. So, a tip of the old horned Viking helmet hat to you, Norway. Jubel!

Iceland, Sweden, Denmark, Finland, the Netherlands

Norway's Northern European brethren aren't too far behind in EV adoption, and all countries have robust government incentives and deadlines to help citizens make the switch.

Again, leadership plays a key role. Iceland, for example, wants bragging rights as the world's first carbon neutral country, and naturally EVs have a key role towards reaching this objective.

In March of 2022, over half of all vehicles sold (56 percent to be exact) were plug in electric, and one of every three new vehicles were electric only. (See Figure 14-2.) Lawmakers have also re-committed to funding purchase incentives, so the trend shows no signs of slowing down.

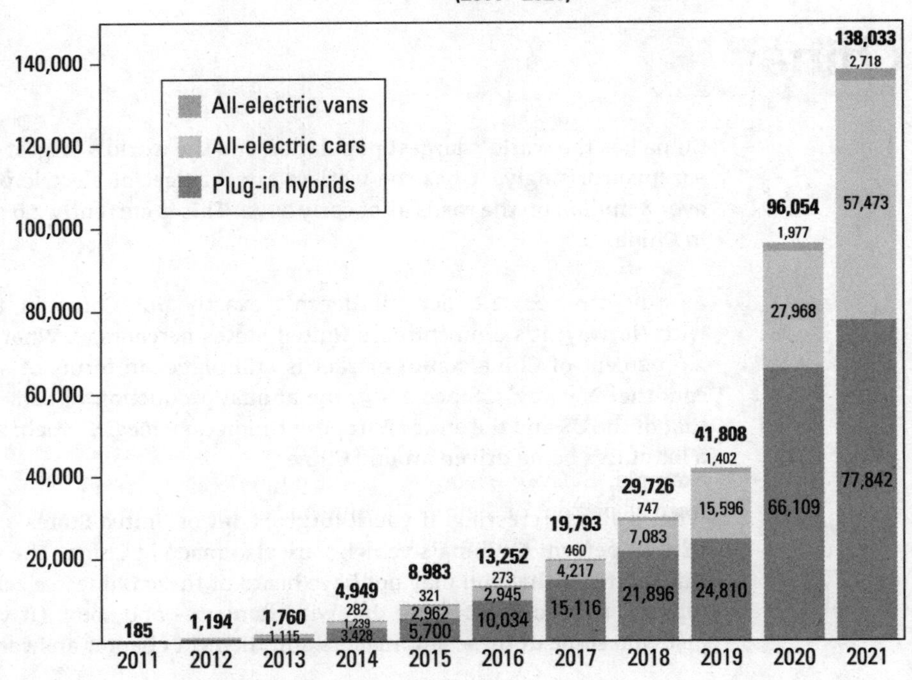

FIGURE 14-2: Annual plug-in sales in Sweden.

Credit: Mario Roberto Duran Cruz. Creative Commons license `https://creativecommons.org/licenses/by-sa/4.0/`

Caption This is “S-Curve” in Swedish, I believe. But do keep in mind that I don’t speak a word of Swedish other than Skal, and I know that word only because it’s written on a shot glass in my cabinet.

Germany

Germany may not be leading Europe in terms of percentages — still at about 2.5 percent of all passenger cars at the time of this writing — but it’s the European leader in terms of sheer numbers, with about 1.2 million plug-in cars registered as of January 2022.

In 2020, the German government approved an economic recovery plan that included direct subsidies for electric cars, which was later raised to €9,000 (approximately $9,650 US), the highest incentive for any European country. The tax incentive applies to both battery electric vehicles (BEVs) and plug-in hybrid electric vehicles (PHEVs), but only for cars costing less than €40,000 ($43,000).

China

China has the world’s largest population and the world’s largest auto market, and so, unsurprisingly, it has the world’s largest fleet of electric passenger cars — over 8 million on the roads as of early 2022. This is currently 2.6 percent of all cars in China.

As you can see, 2.6 percent doesn’t exactly put China in the same league with Norway; it’s a much more United States percentage. What’s notable is that 2.6 percent of China’s auto market is still bigger in terms of sheer numbers of cars that Norway’s. Since 2009, the annual production of cars in China exceeds that of the US and the entire European Union *combined*. So yeah. 2.6 percent is still a lot of EVs being driven around China.

What’s also interesting, if you’ll forgive a bit of United States-centric myopia, is that 98 percent of China’s vehicles are also made in China. The reason I bring up this matter is that you may not have heard of these Chinese electric vehicle automakers, though I’m betting this will change — and soon. (If you’re wondering just who some of these automakers are, the next chapter answers that question.)

Also of note is that China has cut subsidies for electric vehicles by 30 percent in 2022, and it's scheduled to eliminate them completely by year's end. The Chinese government has set a target of having electric vehicles make up a rather paltry 20 percent of new car sales by 2025. Hopefully, the market will make this target seem quaint by then, with more Chinese citizens choosing electric vehicles simply because they offer a better transportation option.

United States

As with China and Germany, the sheer size of the US automobile market makes it impossible to ignore. Also impossible to ignore are the manufacturers headquartered there — Ford, GM, Lucid, Rivian, and, of course, Tesla.

To encourage the adoption of electric vehicles, the US has instituted a federal tax credit of up to $7,500, although that credit has since expired for both Tesla and GM. Alas, subsequent expansion of this popular legislation has been killed off by US politicians, many of whom are beholden to the fossil fuel interests that seemingly fund their lifestyles, which is not how any of this is supposed to work.

In any event, about 5 percent of the cars in the US can now be plugged in (3.5 percent BEV, 1.5 percent PHEV). It's not Norway, but it's not nothing — over 2.3 million as we left 2021 in our rearview mirror. (See what I did there?)

California

If California were a sovereign country, it would have the world's fifth-largest economy — larger than the UK and larger than India. So California's buying habits matter.

In California, about 15 percent of new passenger-car sales are electric and, in 2021, the state saw an 80 percent growth rate for cars that include a charging cable rather than a gas tank.

And, as much as the people in flyover country may delight in pretending that Californians are out of touch, the rest of the US almost always follows California's lead, especially in matters that impact the economy. Generally speaking, if you want to manufacture something meant for global consumption, you make sure it conforms to US law. If you want to manufacture something for sale in the US, you make sure it conforms to California law.

As of early 2022, almost 15 percent of all new car sales in California were battery-only EVs, and the overwhelming majority of those cars were manufactured right there in California.

California has introduced legislation that would ban the sale of gasoline-powered cars by 2035. If that law passes, it's effectively a national mandate because of what I mentioned two paragraphs ago.

UK

The British government has a grant program that allows for a 35 percent discount (max of £3000, or about $3,700 US) on new electric vehicles under $48,670. This is a reduction from a previous grant of $4,170 on vehicles worth up to $69,500, but still.

The UK also has incentives for workplaces that install EV chargers that cover many of the upfront costs of EV charger purchase and installation.

All EV policies and incentives are monitored by an official Office for Zero Emission Vehicles and the country is pursuing an electrification strategy called the Road to Zero. (https://assets.publishing.service.gov.uk/government/uploads/system/uploads/attachment_data/file/739460/road-to-zero.pdf).

As with many countries across Europe, EV registrations began to take off starting around 2014, and demand has only increased since then. Between 2021 and 2022, sales of battery-only EVs are up almost 200 percent (see Figure 14-3).

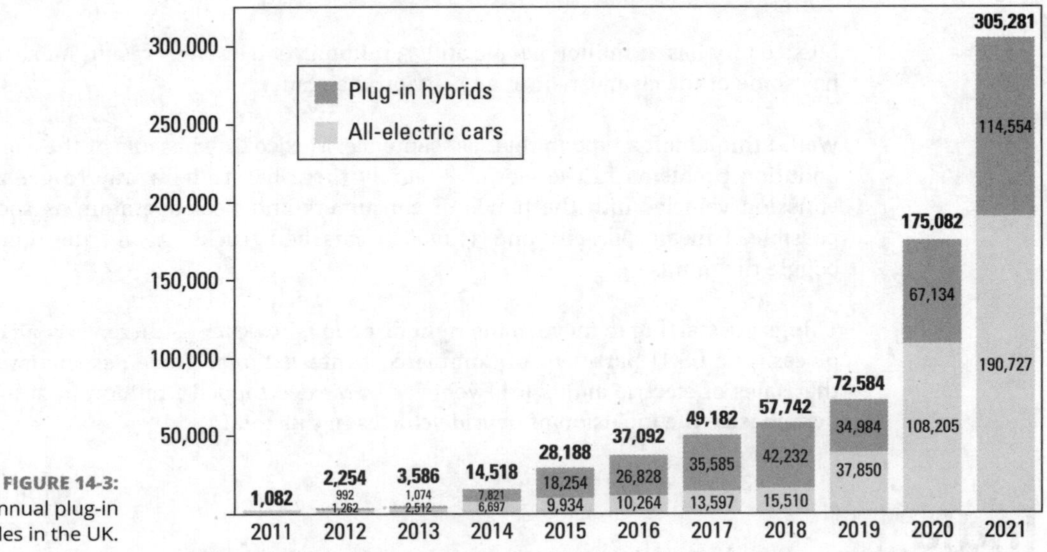

FIGURE 14-3: Annual plug-in sales in the UK.

Credit: Mario Roberto Duran Cruz. Creative Commons license https://creativecommons.org/licenses/by-sa/4.0/

New Zealand

New Zealanders can take advantage of the Clean Car Discount of about US$6,000, which went into effect in July 2021. What's more, the Clean Car Discount also levies extra fees to vehicles that produce high amounts of CO_2, something that's eminently sensible to any parent who has told their kid to clean up their toys.

Note: Many of the facts and figures in this chapter came from a torturously long Wikipedia page. I tend to avoid citing Wikipedia because it's not a first-party information source, but in this case, the article is a good aggregate of several first-party data sources, all of which are carefully cited. Here's that page, if you're curious, but be warned — it's about as long as this entire book: `https://en.wikipedia.org/wiki/Electric_car_use_by_country`.

Mexico

Mexico City has 21 million people and 33 million vehicles. As a result, Mexico City has some of the cleanest air in any city on the planet.

Wait. I think I left a typo in that last sentence. Mexico City has one of the *worst* air pollution problems in the world — surely there has to be a way to get zero-emission vehicles into the hands of consumers and fleet operators as soon as possible. I mean, *one* city and *33 million* cars and trucks. Again, the numbers boggle the mind.

Things are starting to move in the right direction, however — there, as in all other places. The US Department of Commerce's International Trade Association says that sales of electric and hybrid vehicles in Mexico topped 1 million in 2019, but I emphasize the inclusion of hybrid vehicles in this total.

India

The Indian government has recently adopted something called the Faster Adoption and Manufacturing of Hybrid and Electric vehicles (FAME) scheme to incentivize for electric vehicle purchases. Thus far, however, India is very much behind the curve of electric vehicle adoption in a region that really should be leading the world in terms of combining clean energy and zero-emission transportation.

Japan

Japan's electric vehicle uptake of about 1 percent follows the lead of Japan's auto industry, which doesn't seem to believe that electric vehicles are worth pursuing in any meaningful way, a stance that spells the bankruptcy of at least one of Japan's major automakers — Honda or Toyota. (Again, I'm taking the first $1,000 charitable bet if someone wants to take the other side.)

On the other hand, Japan has bullet trains, which should *absolutely* be crisscrossing the United States by now, providing clean, fast, convenient transit to millions.

» Focusing on Asia

» Adding German firms to the mix

Chapter 15
Ten Companies Leading the Way — and a Look at What They're Doing

Here's a rundown of ten manufacturers leading the — okay, I'll just say it outright: These companies are leading the <ahem> *race* for electric vehicle market share.

What? You didn't think I would say "leading the charge," did you? Come on — am I the kind of author who squeezes in a lame electricity pun at every opportunity? I mean, I *am* that kind of author. I've just exercised some very Ohm-ish *resistance* to the temptation in this case.

Boom! That's the electricity pun I was working toward!

At any rate, these are the companies that, in my view, will make most of the new vehicles created in the years to come. (I've added a small amount of conjecture, although the list isn't all that subjective; what's here reflects the current data.)

Tesla

Obviously. It's the world's largest electric vehicle automaker, making about a million vehicles per year before the recent opening of factories in both Berlin and Texas.

Tesla's Model 3 is the highest-selling EV ever, although that baton may be passed to the Model Y when this book goes to print. Tesla also makes the Model X, Model S, Roadster, Cybertruck, and Semi.

It's also the largest automaker in the world *when measured by market cap*, or the number of public shares multiplied by the price per share.

These two factors mean that there's not much to say about Tesla that hasn't previously been mentioned in this book, much less in every third article on the Internet. I'll keep most of the list here focused on makers and models with which most folks are less familiar.

Xpeng Motors

The manufacturer XPeng is headquartered in Guangzhou, China, but also has offices in Mountain View, California (home of Google). It is one of China's fastest growing electric vehicle startups.

In March 2022, Xpeng announced that its sales had increased 202 percent year over year. It currently makes four electric models:

- P5 sedan
- P7 sedan
- G3i SUV
- G9 SUV

The company sells in China, Europe, and Hong Kong.

Lucid and Rivian

I'm lumping together the two start-ups Lucid and Rivian because both are based in California and both had debut vehicles that have impressed the automotive press. Lucid Motors is headquartered in Newark, California, and its debut sedan, the Air, was named *Motor Trend* 2022 Car of the Year. Rivian is based in Irvine, California, and its R1T was named *Motor Trend* 2022 Truck of the Year.

Lucid's assembly plant is in Casa Grande, Arizona, and Lucid is opening a second manufacturing facility in the Kingdom of Saudi Arabia.

Rivian's main manufacturing facility is in Normal, Illinois. The company has plans to build a second major plant in Georgia. Here's a list of their models:

- **Lucid's cars:** The Air sedan and the Gravity SUV (2024)
- **Rivian's trucks:** The R1T pickup, the R1S SUV, and electric delivery vans for Amazon

Volkswagen

Headquartered in Wolfsburg, Germany, Volkswagen is part of the Volkswagen Group, which in turn is one of the two largest carmakers in the world in terms of revenue (Volkswagen and Toyota have been going back and forth the past few years), controlling subsidiaries like Porsche, Audi, and Skoda.

Volkswagen began releasing its ID line of electric vehicles in 2019 and has committed to aggressively expanding its electric offerings over the coming years. in terms of electrics, Volkswagen now sells:

- ID.3 hatchback
- ID.4 Crossover SUV
- ID.5 Coupe SUV
- ID.6 Crossover SUV

The company intends to release the ID. Buzz sometime later in 2022.

GM and Ford

GM and Ford are the old guard of the American auto manufacturing world, which makes them the old guard when it comes to auto manufacturing, period.

Like Volkswagen, however, after having made millions and millions of ICE vehicles over the past 50-100 years, both now have product road maps that include many more EVs in the years between now and 2030.

- **Ford:** Offers the Mach-E hatchback, with plans to begin selling the F-150 electric pickup truck later in 2022
- **GM:** Offers the Bolt hatchback, with plans to begin selling the electric Silverado pickup truck later in 2022

Geeley/Polestar

Geeley, headquartered in Hangzhou, Zhejiang, is the holding company for several auto manufacturers, including Sweden's Volvo and Polestar, Britain's Lotus, and China's Geeley.

Combined, Geeley is the seventh-largest auto manufacturer in China.

Significant electric models under the Geeley umbrella of companies include:

- Polestar 1 (a plug-in hybrid)
- Polestar 2 (a pure-battery EV)
- Volvo XC40 Recharge
- Volvo C40 Recharge
- Geeley Tugella SUV

Most of these vehicles are made in Luqiao, China. The Volvos are also manufactured in Ghent, Belgium.

Hyundai

Based in Korea, Hyundai is the fifth-largest electric vehicle maker in the world. Hyundai's global share of the electric vehicle market sits at about 5 percent at the time of this writing.

Hyundai started its entry in the electric vehicle space with the Kona SUV and has since invested heavily in the Ioniq line, the current iterations of which are the Ioniq5.

The Ioniq investment seems to be paying big dividends thus far, with the 5 receiving a trifecta of awards at the New York International Auto Show: The World Electric Car of the Year, World Car of the Year, and World Car Design of the Year awards in 2022.

I tend to agree: It's a sharp looking vehicle. (See Figure 15-1.)

FIGURE 15-1: The Ioniq5, perfect for a Sharp Dressed Man.

Mike Mareen / Adobe Stock

BYD

The second-biggest Chinese automaker is BYD, which also happens to be the fourth-largest electric vehicle manufacturer in the world, handling between 6 and 7 percent of global electric vehicle sales. BYD is also making significant inroads into the Indian market, and has recently announced a manufacturing facility for electric buses in Lancaster, California

BYD makes a wide variety of vehicles, from small hatchbacks to giant battery-powered trucks.

Check out the options at `https://en.byd.com/bus`.

SIAC

Because SIAC Motor products aren't sold in the US, not many car shoppers there realize the company's significance.

The significance is that it's the second-largest electric vehicle maker in the world, selling about 15 percent of all electric-only vehicles. Although Tesla is gaining ground, SIAC remains the most popular brand in China; in 2021, the Wuling Mini EV city car was the best-selling electric car in the country.

And as we learned in the previous Part of Ten, having a significant share of the Chinese market for EVs means you'll have a large share of all EVs, full stop.

Everyone Else

In this author's opinion, everyone else is playing catch-up: BMW, Mercedes, Nio, Toyota, Honda, Chrysler-Dodge (which is a subsidiary of Dutch manufacturer Stellantis), and more start-ups than I can count.

Some of the nameplates have already introduced fantastic electric vehicles — BMW's i3 (discontinued) and Fiat's e500 (also discontinued) spring to mind. They're just no longer leading the way.

Some of these established players are now beginning to bring exciting battery-electric models to market with a bit more urgency. Some, like Stellantis Chrysler, have announced plans for an all-electric future.

Some of the startups, like Dyson (yes, the vacuum maker Dyson) Sony, (the vision-S concept car), and Fisker (the Fisker Ocean) are either very unlikely to catch up or will never catch up, period.

Even Apple computer has long been at the electric car, working on a "Project Titan" since at least 2018, according to reports. Despite having more cash to throw at an automobile project than any other company in the world, we are still awaiting Tim Cook's announcement.

The fact is that the major automakers left out of this chapter's list were left off for a reason: None has cracked the top five in terms of electric vehicle market share.

Given the size of some of these automakers (the list includes the largest-volume carmaker on the planet) and given the inevitable shift towards electrification that'll happen over the next several years, continued omission from the list above won't make for halcyon days.

They call it disruption for a reason. If you're a shareholder in one of these companies, you have my sympathies.

Chapter 16
Ten Predictions from EV Insiders (with Plenty of Commentary and Opinion Because It's My Book There, Bub)

Scott Galloway, a professor at NYU's Stern School of Business and the host of *The Pivot* podcast, has correctly observed that anyone in the prediction business is basically peddling the equivalent of horse manure.

But I'm here to say (and no horse droppings this time) that predictions are fun!

The penalty for getting them wrong is minimal for people like me. I'm not paid to get the prediction business right, like a baseball talent scout or a meteorologist on the local news. In that way, I am to the EV prediction business what Jim Cramer is to the stock market business — just someone running my mouth. We get most of the predictions wrong, but the paycheck clears either way.

But still — lots of fun! And also, valuable! As Galloway says, again correctly in my opinion, "The value of a prediction is in the act of making it, not the prediction itself. Contemplating what may happen encourages us to take responsibility for decisions we make in the present."

So, with that responsibility in mind — about the decisions we might need to make in the here and now — here goes.

When New EVs Will Cost $25,000

The price of EVs – or at least a model of Tesla EV – will hit $25,000 in 2023 — or at least in 2023 according to Elon Musk in 2021. You can read more about that statement here: `https://electrek.co/2021/09/02/tesla-aims-to-release-25000-electric-car-in-2023-likely-will-not-have-a-steering-wheel`.

Musk is an EV insider, so it doesn't really matter what you think of his track record for predictions. I think even he would agree that tend to be — how do I say this? — *optimistic.*

The matter at hand is the prediction itself. My response is this: it's *possible*, yet the fundamental problem with that forecast is that electric vehicle manufacturers — Ford, Hyundai, Lucid, and others — have little incentive to make a $25,000 electric vehicle when they're selling all the ones they can make for twice that amount or more. The Ford F-150 pickup has a three-*year* waiting list as I write this, and I would bet my annual salary that there'll never be a $25,000 F-150.

Manufacturers will make new $25,000 electric vehicles when they're forced to because of market conditions, which in my mind means never. I would just cross that price off the wish list, if I were you. You have plenty of *used* options at that price point, though.

Update: between the drafting of this very first prediction and the revision, Chevy lowered process on its 2023 Bolt EUV by almost $6,000 when compared to its 2022 model, making a brand new Bolt EV available at $26,595.

And you can get a Kandi K23 in the United States for the no haggle price of $14,999. (See Figure 16-1.) So Musk's prediction about price was accurate after all, even of the company making it was not.

FIGURE 16-1: I want Kandi! (The Kandi K23).

Kandi America

When Robotaxis Will Make Their Grand Entrance

The world has been promised robotaxis for many years now, and the loudest voice has been that of Elon Musk, who has been promising robotaxis "next year" since 2016. But you know what? Robotaxis are here *now* — in 2022! You can ride in a robotaxi today in San Francisco or Phoenix. Just look for a Waymo (Google's self-driving outfit, discussed in Chapter 11) when you land at Sky Harbor Airport and it will ferry you around without a safety driver.

My two cents on the subject of robotaxis is that humans lose sight of just how quickly 50 years go by, and sometime between today and 50 years from now, humans will do for driving what they've done for computers, medicine, transportation, and other technologies.

For better or worse, humans are a species of *very* smart apes — the smartest entity in the known universe, in fact. So I try to keep the ling view in mind when it comes to these sorts of predictions. After all, in 5 billion years, we'll have to solved for what to do when the sun becomes a red giant and vaporizes the entire planet. We'll surely have fully cracked to code of robotaxis by then.

And between then (the full robotaxi revolution, not the Earth getting swallowed by the sun) and now, there are companies like Halo that bridge the gap. Launched in 2022 in Las Vegas, Halo will remotely drive an electric car to you, let you then drive it to your destination, and then let you walk away, where it will be remote controlled once more to the next customer (or to a garage where it can charge). For those curious, visit `https://halo.car/` for more info.

When Electric Planes (or Flying Cars) Will Make *Their* Grand Entrance

It's only natural to extend the conversation about electric-powered automobiles to electric-powered flight. Aircraft makes up roughly 10 percent of overall transportation emissions, so it's not as significant to the zero-emission equation as the automobile market, but every little bit counts.

With that said, it looks like *passenger class* electric airplanes may be arriving as soon as mid-decade. A good overview for your next (first?) lazy Saturday reading list can be found at `www.afar.com/magazine/electric-planes-are-coming-sooner-than-you-think`.

Of course, Denmark and Sweden have pledged to make all domestic flights free of fossil fuels by 2030. Can I just move to Denmark now? If you're in Denmark and you're hiring, you know how to reach me.

As you might guess, most of the power is used at takeoff, and then the plane just needs a relatively small amount of power to keep itself aloft. And, on the way down, the plane can actually recapture much of that energy, just like a car can do by using regenerative braking.

Beyond that, I've already shared my thoughts on flying cars in Chapter 5.

Which Companies Will Thrive (and Which Ones Won't)

Entire books will undoubtedly be written on the subject of thriving companies before too long (maybe I'll write one — again, you know how to reach me), so what I can do here is present my predictions about the next five to ten years, and then we can all look back in 2025 to see how foolish I sounded.

I'll keep each list short. Here goes:

Electric vehicle or EV-related companies that will thrive

- BYD
- Hyundai
- Tesla
- Volkswagen
- Xpeng

Electric vehicle or EV-related companies that will likely go bankrupt by 2030

- **General Motors:** GM makes the list mostly because they have taken large investment stakes in companies like Lordstown and Nikola while also having a track record of making and destroying what could have been a game-changer of a vehicle: the EV1. GM now talks a fairly good electric vehicle game, but in my opinion, GM shareholders and employees would be better off by liquidating the entire company *today* and investing in EV makers like Hyundai or Tesla or Xpeng.
- **Nikola Motors:** Nikola has little credibility and little chance of surviving a in a market that is starting to see deliveries by the likes of Daimler and Volvo and Kenworth.
- **Nio:** Nio seems to have a great product lineup
- **Toyota:** This is the most bold of my predictions, and the one that has the most risk of making me look like a true Dummy; it's also the company that I think looks most like Blockbuster or Blackberry or Kodak in terms of its product lineup. Yes, I know they sell a lot of plug-in hybrid RAV 4s. But Kodak sold a lot of film. I just think Toyota has bet their future on a technology that doesn't have one.

Electric vehicle or EV-related companies that haven't yet launched (or won't ever launch) production

- Byton (launched in 2018, bankrupt in 2021)
- Dyson (launched in 2016, cancelled in 2021)
- Fisker (bankrupt in 2014, relaunched in 2020 via a Special Purpose Acquisition Company (SPAC), announced deliveries of the Ocean in 2021, then early 2022, then production beginning in November 2022. . . I'm sensing a pattern here.)
- Chanje (bankrupt in 2021) I could devote an entire chapter to Chanje, and how the company intersects with a (since closed) factory in Kansas City of all places, and a meeting with President Obama, a story that's most definitely worth your time if you're as fascinated with this as I am. What space allows for instead is a quick summary of Chanje's origin, which revolves around CEO Bryan Hansel's trip to the Amazon rainforest sometime in 2015. Some of that vacation time was spent tripping on ayahuasca, during which he received a message from mother Gaia. The trip(s) were transformative for Hansel, so much so that when he returned, he started an electric delivery truck company that would import Chinese vehicles and sell them to big delivery companies like FedEx, to whom it promised delivery of 1,000 trucks. He also reportedly encouraged his employees to take ayahuasca and to meet meditation quotas. I know what you're thinking. Electric cars? Ayahuasca? What's not to like here?

  Unfortunately, the vision outpaced the ability to deliver. The company folded in 2021 when the Chinese firm that Chanje was counting on to manufacture the electric trucks it had promised to Ryder and FedEx went bankrupt itself.
- Faraday Future (founded in 2014, went public via a SPAC in 2021, under investigation by the SEC and also the Department of Justice in 2022 for, among other alleged misconduct by executives, overstating its reservation numbers by, oh, about 99 percent, "plans" to open a factory in China in 2025. *Future* indeed.)
- Lordstown Motors (founded in 2018, went public in 2020 via a SPAC — these are different companies, I swear, not variants in a Marvel Multiverse — and bought the former GM assembly plant in Lordstown, Ohio. They also have paid the Workhouse Group (a legit truck maker) for *licensing* rights to Workhorse's electric pickup pick, for which they started collecting reservations. Under investigation by the SEC in 2021 for . . . wait for it . . . overstating reservation numbers! Did I mention that the founder of Lordstown is a former exec at Workhorse? Apparently, there's a playbook for this sort of thing.)

As this book neared its deadline, Lordstown sold off its namesake plant to Taiwan's Foxconn for $230 million, which saved the truck "maker" from closing its doors outright. What's Foxconn going to do with the factory, by the way? Reportedly, they're going to build electric cars for clients like <checks notes> Lordstown and Fisker. Fisker deliveries to begin in 2024. Okaaay.

Don't worry, though: Whether these two companies ever make cars, the executives have purchased second and third homes in places like Miami Beach or The Maldives, I suppose, because capitalism is gonna capitalism.

The takeaway? Making lists is easy. Making cars is hard.

When You Will See Electric Semis (and What It Will Mean)

Thanks to companies like Volvo, BYD, Daimler, and Kenworth, the answer to that is right now.

Beyond that, this answer has a bit of nuance to it — hey, you should check out Chapter 12, which covers this in more detail! — because while there are electric semis currently roaming a few of the nation's roads, most are still in the prototype stage, and none are yet ready for the long-haul trucking duties that remain the purview of the diesel 18-wheeler.

It's worth remembering that Tesla, a very successful electric automobile maker, has still not solved the riddle of electric long-haul trucking. Tesla debuted its semi called Semi in 2017. (The Tesla creative team wasn't on Twitter at the time, apparently, and didn't respond to the "what's a good name for this truck thingie we're making?" tweet.) Anyway, at the time of the announcement, the release date was scheduled for 2020. The most recent claim by Elon, made in April of 2022, is that production will start in 2023. (No word on whether Tesla's creative team is now on Twitter.)

Meanwhile, another startup trying to make electric semi trucks was founded and taken public, and then later found its CEO the subject of a fraud investigation surrounding his claims about the company's ability to manufacture an all-electric long-haul semi. That company was Nikola, and as I write this chapter, it does appear that the company is ready to begin production of its Class 8 truck, which goes by the name Tre.

Bottom line? 2022 is when the world will see the first battery electric semis making their first deliveries – hopefully filled with giant palettes of *EVs for Dummies*.

Future Battery Tech for Your Car

If by *future battery tech* you mean solid state batteries and by *car* you mean a plug-in hybrid, then the date for that event looks to be 2025, when Toyota says it will produce a PHEV using a solid-state battery. This info comes from an interview with Toyota's chief scientist while at the 2022 Consumer Electronics Show in Las Vegas.

The batteries used for electric vehicles produced in 2030 will be quite different (in particular, in their chemistry and their cathode-anode composition) from the lithium ion batteries prevalent in present-day electric vehicles. I'm especially interested in following developments in the lithium-sulfur space.

When You'll (Hopefully) See an Impact on Parking and City Design

Los Angeles, arguably the most car-centric city on the planet (though I'll listen to arguments for Phoenix, Las Vegas, or Houston, all of which are depressing hellscapes as a result), *has more acreage dedicated to parking cars than there is land in all of Manhattan*, in New York. If you're keeping track, 1.7 million people live in Manhattan.

If you're counting square footage rather than people, then consider that LA uses 27 square miles for parking — and then another 200 square miles of parking space in LA county proper. Think about what a staggering waste of space and money has been handed over from the public realm so that private, single-passenger autos can bake in the sun for hours on end.

Here, see for yourself: `www.betterinstitutions.com/blog/2016/1/2/map-a-parking-lot-with-all-of-la-countys-186-million-parking-spaces`.

Now imagine what life might be like with autonomous vehicles that never need to park. Imagine life if all that space — heck, imaging even if one fifth of that space — were used for parks. Or for urban farming. Or for housing, of course. And what if much of that housing were located near enough to employment centers so that even more people could walk or bike to work?

It's an exciting future to contemplate. And I think that, eventually, it will come to pass — truly autonomous driving vehicles by 2030, and the initial effects of redeveloped parking lots ten years after that.

From an economic and societal standpoint, continued sprawl is unsustainable. The autonomous car can be the catalyst that helps much of humanity reverse this ruinous course.

Connected Cars

The concept of *connected* cars can have different meanings, depending on the context, but here I'm referring to a car's ability to "talk" to other cars, communicating info such as speed, location, and direction — all of which can help them travel more safely. (The other meaning of a connected car is that it's connected to the Internet, which essentially is what happens already with most electric vehicles. If you can use an app to tell your car to stop charging or turn on the seat heaters, for example, then in this context your car is already connected.)

The advantage of a connected car network is that the vehicles would be autonomous. They would never crash. They would never experience road rage. They would rarely be subjected to traffic jams, which are caused in large part by cars competing for space.

Because vehicle-to-vehicle communication has the potential to bring numerous societal benefits, the US Department of Transportation (DoT) has been making strides in this area for the past several years. The fly in the ointment lies in all the cars that aren't (yet) networked. Getting cars to communicate is now a relatively easy engineering task. One of the big questions that remains unresolved is how to design a system that gradually integrates vehicles that can talk to each other with vehicles that cannot.

The other big question is how to integrate such a system with human not in a car. There's a pair of excellent point-counterpoint YouTube essay on this entire subject if you want to spend an extra 20 minutes contemplating the idea of a connected car. Creator GCP Grey has a "*Simple Solution to Traffic*" which presents the point (source: `https://www.youtube.com/watch?v=iHzzSao6ypE`. It's been viewed 30 million times, so people are obviously intrigued by this vision of automobile utopia), while the counterpoint is from a creator called Adam Something, and the video is "*The ACTUAL Solution to Traffic - A Response to CGP Grey*" (source: `https://www.youtube.com/watch?v=oafm733nI6U`).

I tend to side with Adam Something side of this discussion.

I'll wrap things up with a risk-free estimate of 2035-ish, and even then the system will be rolled out in a geo-fenced fashion. (A geo-fence refers to a virtual boundary around a real geographic area, one that's set by predefined GPS coordinates. Geo-fencing requires a device that's aware of its location, like a phone or like almost any electric vehicle.)

The Last ICE Sale Will Take Place Sometime in 2030 — for California

Loren McDonald, an EV insider of the website EV Adoption, says, "When looking at the state of California, our 2030 forecast for just BEVs (not including PHEVs) shows BEVs as a share of new vehicle sales reaching nearly 57 percent."

(The site, at `https://evadoption.com`, provides data analysis and forecasts to industry observers and organizations that are helping drive the transition to electric vehicles.)

REMEMBER

California is scheduled to require that all new cars and passenger trucks sold to be zero-emission by 2035. I would bet anyone willing to take the other side of a $1,000 bet — donated to a charity, of course — that California will move up the date. The bet is that no rational Californian will buy a new ICE starting in 2030.

Before you email me saying you want to take the bet, I'll leave you with Figure 16-2.

FIGURE 16-2: Do you really want to pay this much?

MichaelVi/Adobe Stock

Index

Numbers

A

B

C

D

E

F

G

H

I

J

K

L

M

N

O

P

R

S

T

U

V

W

X

Z

About the Author

Brian Culp rode to dinner in an electric vehicle in 2017, and the experience changed his life. He became an EV owner in 2018, an EV automaker employee in 2019, and most recently, an electric bicycle owner in 2021.

As an EV owner and enthusiast, he's spent almost every spare moment between 2018 and today learning about charging infrastructure, battery chemistry, and the engineering (and ethical) challenges surrounding autonomous driving. As the Technical Publications manager at Lucid Motors, Brian has gotten first-hand knowledge about EV tech, performance, engineering, and service.

Additionally, Brian has an extensive track record of explaining technical concepts to a non-technical audience. He's presented at MacWorld and for Microsoft Across America. His courses on Udemy are global bestsellers. And, over a 15-year span, he's authored or contributed to over 30 publications.

Brian and his wife (who is a software engineer for a competing EV manufacturer) live in Silicon Valley. He loves hearing from readers and students, and you can connect at linkedin.com/in/hmsbrian, or email directly at `hmsbrian@gmail`.

For those wondering: I set Kansas State's all-time home run record with the last swing I ever took in college during the Big 8 baseball tournament. Yes, there was a conference called the Big 8 back then. For someone who mentioned it on the cover of a book, I'm actually far less proud about it than it might seem. Yes, baseball was fun, and yes, I'd probably do it again, but ultimately hitting a baseball has no impact on the world. I guess certain moments could have made my family and teammates happy for a fleeting moment. (Other moments almost certainly gave them heartburn.)

I'm starting to get philosophical about it, I guess, which is what old men do when they gain perspective. I think the word is "wisdom" or "humility"; in my case, these are very much works in progress. What I'm saying by way of this mild lament is that playing baseball seemed to matter at the time. It didn't.

EVs matter. The planet matters. If I'd seen the world then as I see it now, I'd not spend all that time and brainpower on baseball. I'd have learned more about carbon sequestration. Or battery engineering. Or generating electricity using solar and wind.

Failing the invention of time travel, I've done my best to learn from those who did learn about these things in school, rather than how to hit sliders left up in the strike zone.

Dedication

This book is dedicated to my daughter, who finds herself coming of age on a planet that is inexorably warming. She, along with billions of fellow humans and uncountable numbers of the planet's flora/fauna, will have to adapt to the effects of the Anthropocene despite having no say over its many causes.

I hope that electric cars are part of an effort to change our current trajectory. In turn, I hope this book is a (very) small part in someone's decision to stop buying and burning gasoline. No, it won't matter much. But it will matter. Like I say below, you never know who's watching; you never know who will pick up where you left off.

Author Acknowledgements

Creating an electric vehicle is a team effort. So it is with a book of this size and scope. Many, many people had a hand in creating the thing you now hold and trust me when I say that the finished result is better for it. Any errors, omissions, or overwrought prose are the byproduct of your author's limited writing skill, knowledge, time, or any combination thereof.

Some of the people who were part of this team effort deserve a specific mention:

Vicki Adang, who shepherded me through the "pilot" chapter, and helped me execute on the unique *For Dummies* style. Her work during the book's incubation stage set the book up for success.

Paul Levesque, who served as the book's development editor and project manager, keeping things on time and on topic (as you may have noticed while reading, I'm not above a bit of parenthetical self-indulgence — I mean jokes or asides that deeply enrich the reading experience). Paul also served as the graphics and clearance editor, although he did not know he was signing on for that role at the time. What I'm saying is that Paul worked his behind off getting this book out, and that work did not go unnoticed.

Steve Hayes, Steve was the book's acquisitions editor, which means that Steve is the person who took a chance on this topic and then entrusted me to be the caretaker for that topic. The best way to repay that trust is to deliver an interesting, entertaining read. I hope that's the case.

Carole McClendon, I've had the great fortune of having Carole as an advocate and strategist over the past 15 years or so. Here's to 15 more years of doing what we love.

Floyd Smith, Many thanks for your input and research into the topic of autonomous driving, and for your generous pointers about writing in the *For Dummies* style. (Floyd, by the way, is the author of the deservedly bestselling *Creating Web Pages for Dummies*.)

I also want to acknowledge the work of people who've had a hand in my ability to create a book on this topic, even though they don't know it and wouldn't know me from a Z-man sandwich from Joe's Kansas City BBQ, which is a sandwich everyone should have once in their life. (Joe's ships in case you're wondering.)

I mention in the book's Intro that discovering electric vehicles was a brain-on-fire moment for me in 2017, and that feeling of excitement about the possibilities of electric vehicles has not abated to this day. The futurists and tinkerers and thinkers below played a key role in igniting that curiosity, and you might notice that several of them don't even spend much time talking about cars. In no particular order, they are:

Charles Mahron (author of *Strong Towns: A Bottom-Up Revolution to Rebuild American Prosperity*), Jason Slaughter (The *Not Just Bikes* YouTuber), Rollie Williams (*The Climate Town* YouTuber), Tony Seba (the force behind the RethinkX think tank), Ramez Naam (co-chair for Energy and Environment at Singularity University), Kyle Conner (founder of Out of Spec Studios), Rob Maurer (host of the Tesla Daily YouTube channel), Matt Ferrell (The *Undecided with Matt Farell* Youtuber), Ricky Roy (host and CEO of the Two Bit da Vinci, YouTube channel).

That list could probably go on for another page, but the one lesson I learned from these folks (perhaps more than any other) was this: Do good work. Show it to people. Most everything else is luck and/or out of your control, but you never know how far the ripples go when tossing a pebble of creativity into a pond of ideas. I can think of no greater honor for this work than to have it randomly cited by someone who may perhaps be inspired by this.

Finally, I want to acknowledge my family for taking the electric vehicle leap in 2022. My parents and in-laws both got their first EVs this year, as did my brother and his family. Meanwhile, my sister's family installed a solar array on their rooftop and followed it with not one but two Level 2 outlets in their garage. Deciding on the EV that works best for them is the next step. If you read references to families taking kids to hockey tournaments, I was mostly picturing my sister during those moments.

Besides that, the only audience I ever truly write for is my wife. She's read most of what you just did long ago, and she liked it, or at least said she did, which in my eyes makes the book a towering success. Whenever I think I worked hard on this book, I think of the software engineering work Jen does on the daily. Thanks Jen, for the example you set for me, along with many others impacted by your work.

Publisher's Acknowledgments

Acquisitions Editor: Steve Hayes
Senior Project Editor: Paul Levesque
Copy Editor: Becky Whitney
Tech Editor: April Bolduc
Production Editor: Tamilmani Varadharaj
Cover Image: © SimonSkafar/Getty Images

Leverage the power

Dummies is the global leader in the reference category and one of the most trusted and highly regarded brands in the world. No longer just focused on books, customers now have access to the dummies content they need in the format they want. Together we'll craft a solution that engages your customers, stands out from the competition, and helps you meet your goals.

Advertising & Sponsorships

Connect with an engaged audience on a powerful multimedia site, and position your message alongside expert how-to content. Dummies.com is a one-stop shop for free, online information and know-how curated by a team of experts.

- Targeted ads
- Video
- Email Marketing
- Microsites
- Sweepstakes sponsorship

20 MILLION PAGE VIEWS **EVERY SINGLE MONTH**

15 MILLION **UNIQUE** **VISITORS PER MONTH**

43% OF ALL VISITORS ACCESS THE SITE **VIA THEIR MOBILE DEVICES**

700,000 NEWSLETTER **SUBSCRIPTIONS** **TO THE INBOXES OF** *300,000* UNIQUE **INDIVIDUALS EVERY WEEK**

PERSONAL ENRICHMENT

9781119187790
USA $26.00
CAN $31.99
UK £19.99

9781119179030
USA $21.99
CAN $25.99
UK £16.99

9781119293354
USA $24.99
CAN $29.99
UK £17.99

9781119293347
USA $22.99
CAN $27.99
UK £16.99

9781119310068
USA $22.99
CAN $27.99
UK £16.99

9781119235606
USA $24.99
CAN $29.99
UK £17.99

9781119251163
USA $24.99
CAN $29.99
UK £17.99

9781119235491
USA $26.99
CAN $31.99
UK £19.99

9781119279952
USA $24.99
CAN $29.99
UK £17.99

9781119283133
USA $24.99
CAN $29.99
UK £17.99

9781119287117
USA $24.99
CAN $29.99
UK £16.99

9781119130246
USA $22.99
CAN $27.99
UK £16.99

PROFESSIONAL DEVELOPMENT

9781119311041
USA $24.99
CAN $29.99
UK £17.99

9781119255796
USA $39.99
CAN $47.99
UK £27.99

9781119293439
USA $26.99
CAN $31.99
UK £19.99

9781119281467
USA $26.99
CAN $31.99
UK £19.99

9781119280651
USA $29.99
CAN $35.99
UK £21.99

9781119251132
USA $24.99
CAN $29.99
UK £17.99

9781119310563
USA $34.00
CAN $41.99
UK £24.99

9781119181705
USA $29.99
CAN $35.99
UK £21.99

9781119263593
USA $26.99
CAN $31.99
UK £19.99

9781119257769
USA $29.99
CAN $35.99
UK £21.99

9781119293477
USA $26.99
CAN $31.99
UK £19.99

9781119265313
USA $24.99
CAN $29.99
UK £17.99

9781119239314
USA $29.99
CAN $35.99
UK £21.99

9781119293323
USA $29.99
CAN $35.99
UK £21.99

dummies.com

dummies®
A Wiley Brand